QUALITY ASSURANCE IN CONSTRUCTION

Second Edition

QUALITY ASSURANCE IN CONSTRUCTION

Second Edition

Brian Thorpe
Peter Sumner
John Duncan

Gower

First edition published 1990

Gower Publishing Limited
Gower House
Croft Road
Aldershot
Hampshire GU11 3HR
England

Gower
Old Post Road
Brookfield
Vermont 05036
USA

Brian Thorpe, Peter Sumner and John Duncan have asserted their right under the
Copyright, Designs and Patents Act 1988 to be identified as the authors of this work.

British Library Cataloguing in Publication Data

Thorpe, Brian
    Quality assurance in construction. - 2nd edn.
    1. Civil engineering - Quality control 2. Construction
    industry - Quality control
    I. Title II. Sumner, Peter III. Duncan, John M.
    624'.0685

    ISBN 0 566 07758 2

    Library of Congress Cataloging-in-Publication Data

Thorpe, Brian.
    Quality assurance in construction / Brian Thorpe, Peter Sumner,
John Duncan. — 2nd ed.
        p.   cm.
    Duncan's name appears first on the previous ed.
    Includes index.
    ISBN 0–566–07758–2 (hardback)
    1. Engineering—Great Britain—Management.   2. Construction
industry—Great Britain—Quality control.   3. Quality assurance—
Great Britain.   I. Sumner, Peter.   II. Duncan, John M.
III. Title.
TA190.D85   1996
690'.094—dc20
                                                                96–1807
                                                                CIP

Typeset in Palatino by Raven Typesetters, Chester and
printed in Great Britain by Hartnolls Ltd, Bodmin.

Contents

Illustrations

Preface

Since the first edition of this book was published, in 1990, many changes have taken place both within the construction industry and in the field of quality management. As forecast in the previous edition, quality management systems have now been introduced by thousands of discerning companies who rightly saw such systems as a valuable marketing aid and, more importantly, as a vehicle to help them meet their clients' needs completely, whilst simultaneously achieving their own prime business objectives.

During the past five years industry in general has been faced with continually increasing demands concerning aspects such as health and safety (the 1994 CDM regulations being a typical example – see Chapter 9 of this book); environmental issues, both local and global; continually emerging EC regulations and so on.

Quality Assurance similarly has moved toward the more all embracing philosophy of Total Quality Management (TQM), with recognition of the internal customer, the levelling of management hierarchy structures and of the continuing business improvement benefits to be gained by creating an environment and a culture which frees and utilizes the latent experience and knowledge of each individual.

The standard BS 5750 of 1987 has been superseded by BS/EN/ISO 9000 of 1994 to better reflect current management and control requirements. In view of these changes within the marketplace, it was decided that a revised edition of the book would be appropriate.

This new edition therefore, whilst retaining the general format of the original, endeavours to reflect more closely the current situations and to provide a useful aid to those still contemplating going down the quality management route, as well as those who now have well tried and tested management systems.

Since writing the original work, the authors also have changed their occupational roles.

Brian Thorpe, CEng, MIMechE, MIEE, MIED, FIQA

Now retired from full-time quality management employment, Brian continues to write new training material and provides quality management training and consultancy in a private capacity.

Peter Sumner, MIM, DMS, MIQA

Peter is currently working as the Head of the Quality Management Division within a firm of design consultants based in the north of England.

John Duncan, BSc, CEng, FICE

John is currently engaged as a Regional Director with a multi-discipline consulting engineering organization.

Despite these changes however, the writers have again collaborated to prepare this new edition and hope that it will provide an enjoyable read and be of genuine help to readers in their understanding and utilization of quality management techniques.

It should be recognized that a well-structured, properly understood, selectively and sensibly applied quality management system, will improve your marketing capability, competitive edge and business performance, and provide an essential springboard for ongoing improvement.

We hope that this book will help you to achieve these benefits.

Definitions and abbreviations

CDM: Construction (Design and Management) Regulations, 1994

DETAILED QUALITY PLAN (DQP): A statement of the activities, responsibilities, procedures, verifications and records pertaining to the carrying out of a discrete process or sequence of related activities

FIRST PARTY ASSESSMENT: Assessment carried out internally by an organization on its own operations

SECOND PARTY ASSESSMENT: Assessment of a provider of a product or service, by the client/customer

THIRD PARTY ASSESSMENT: Assessment of an organization by a recognized external assessment body or individual on behalf of either a potential customer or the organization being assessed. (At the time of this publication, there are approximately 40 external assessment organizations, known as certification bodies, who have received accreditation from the National Accreditation Council for Certification Bodies to carry out third-party assessments.)

H&S: Health & Safety

H&SE: Health & Safety Executive

PREFERRED SUPPLIERS'/PROVIDERS' LIST: A formal and maintained list, based upon the records of suppliers'/providers' continuing overall ability to plan, resource and meet specified requirements (including those for quality and for health and safety)

QUALITY MANAGEMENT SYSTEM: The organizational structure, responsibilities, processes, procedures and resources for implementing quality management

QUALITY PROGRAMMES/PROJECT QUALITY PLAN (PQP): A description of the overall management, procedures and controls pertaining to the execution of a specific contract, project or stage thereof

TQM: Total Quality Management

1   Introduction

The state of the art

Since the initial publication of this book in 1990, the state of the art in terms of quality assurance has changed significantly.

- Within the UK alone, it is estimated that in excess of 40,000 organizations have now gained third-party certification for their quality management systems. However, we see this as only the tip of the iceberg and our perception is that for every organization which has sought and gained certification, there are at least as many again who have established sound systems to enhance their business performance without yet seeking formal recognition.
- The ability to demonstrate a quality management capability has increasingly become a tender requirement.
- Many organizations who have had their systems established for some time have now moved into the continuous improvement arena, recognizing that the greatest asset of any business are the people who work within it and who have the knowledge and experience to drive improvement.

- Others are aiming at even more all embracing goals such as the achievement of national or international quality awards.
- The new ISO 9000 (1994) standard has now been adopted in almost every industrialized country in the world (including the US and Japan) as a reference for setting up quality management systems, thus creating a recognizable benchmark across international trading boundaries.
- Environmental issues are now rightly an important consideration, whilst the arrival of the 1994 CDM Regulations has added a new dimension concerning health and safety in construction. New EU regulations come into force almost weekly. These aspects all need to be accommodated within the framework of our management and control system.

For those who already have a good quality management system, the vehicle to manage the requirements arising out of the above will, to a large extent, exist. Those who do not yet have such systems may find things somewhat more difficult and costly to manage.

We hope that this book will encourage those who have not yet embarked down the QA path to do so. It is not really difficult, providing a sensible and logical approach such as that described in the following pages is adopted.

To those who already have systems in place, we hope that those sections concerning the selective and cost-effective use of the system will be of value.

The need for quality assurance in construction

Over the years much research has been done concerning the incidence and sources of failures within the construction industry. There is some disagreement as to exactly where the blame should be laid in terms of statistics. However, it is generally recognized that the major sources of cost-related error lie in either the design team or the construction management team, with the prime sources of such errors being seen as:

- inadequate training and management of the designers responsible for producing calculations and drawings. This is realized in excessive changes to the detail information throughout the construction period of the project and, consequently, variations to the construction costs
- inadequate or incorrect specification at tender
- inadequate definition of responsibility within both management groups, that is, in the office and on site
- poor communication between the principal parties in contract, which ultimately leads to confusion and cost-related delays
- inadequate training and management of the technicians and labour on site
- inadequate verification routines to ensure that design, materials and workmanship meet specified requirements.

Why should we expect quality assurance to provide an answer to the problems stated above? Perhaps we need to look a little more closely at the reason why these problems occur.

No matter what sector of the profession we work in, the art of communication is vital to each and every individual. Instructions must be clearly given and understood, calculations must be accurate and working documents clear and unambiguous in their interpretation. The analysis of error in construction very quickly yields one pertinent fact: our inability to communicate effectively is our biggest enemy.

When dealing with a complex and varied industry, within which numerous professionals and artisans operate, whose background training and professional development are entirely different from each other, the most effective way to achieve good communication is to formalize it.

How often we observe in contractual dispute, the architect blaming the engineer blaming the contractor for errors which have occurred simply because drawings were incompatible or an instruction was misinterpreted. Formalized systems adopted by all parties for the checking and recording of communications would eradicate many of the problems.

Training at most levels within the industry has been sadly

neglected for many years. Classic disasters have occurred because of inadequate training of those responsible for key activities. Formal QA requires that training policies for all staff are developed and implemented.

In addition to the importance of good communication and training, a study of contract disputes very often yields a surprising lack of understanding as to exactly who is responsible for the principal decisions in the construction process.

By this we do not mean that there is confusion within conditions of contract, but that the internal management systems of the professional bodies associated with the contract are inadequate. Very often decisions are made by individuals who are entirely unqualified to make them, and in most cases those individuals believe that the ultimate responsibility will be carried by an immediate superior. Formal quality assurance demands that all levels of staff associated with the construction process have a clear definition and understanding of their own limits of responsibility.

In addition, all individuals must understand how they fit into the overall management of the parent organization and relate to their fellow team members.

We believe that the following statement summarizes one of the main problems experienced by those responsible for Quality in Construction: 'In the process of undertaking work within a tight programme there never seems to be sufficient time or money to ensure that the results are correct. However, on discovery of a major error or fault the resources to put it right are limitless.'

Time is the biggest thief of quality. How often it is that a client, or client's agent, will press for a programme that is entirely unrealistic and in many cases unnecessary, which will ultimately lead to disaster. A very common example of this is the abbreviated soils report or, in some instances, the non-existent soils report. There never seems to be enough time or money to ensure that the soil is suitable to withstand the forces and stresses to which it will be subjected. However, this is so often the root cause of a large/cost-related failure.

This question of allowing sufficient time has been recognized as important by the health and safety-related CDM Regulations

1994, which expressly ask that the key resource of adequate time for the proper performing of stipulated requirements be a pre-appointment consideration factor.

The QA approach seeks proper pre-thought about what is to be done, so that planning and resourcing, including time, can be based upon real understanding and then reflected in control documents such as Project and/or Detail Quality Plans.

Anyone who has been associated with the process of litigation within the construction industry is aware of the difficulties in progressing a claim, or indeed the defence of a claim without proper and accurate records. Construction works begin their life on the date that they are commissioned and, for many reasons, clear and accurate records of the construction and design are essential. Quality assurance requires that formal records are kept throughout the period of design and construction as well as being archived following completion.

The very process of identifying and preparing these records and maintaining them from the start of the project acts as a stimulant to the professional and construction teams to ensure that the above stated requirements of good communication, training and definition are carried out.

Many have argued that the quantity of paper produced as a result of formal quality assurance is its biggest enemy. This book makes it clear that the documentation required by a formal QA system is necessary to basic good management, provided that the system has been designed only to accommodate the requirements of the particular company or group. Sensibly applied and streamlined quality management systems are not only beneficial but essential to the construction industry of tomorrow.

2 The Meaning of Quality Assurance

Definition of the terms

Ask a few people to define the word 'quality' and it will soon be apparent that opinions vary very widely indeed. Some people will say 'the best', others 'value for money', whilst some will offer the definition of 'fitness for purpose'.

This latter definition, which appears in many quality-related documents, may be true in the final analysis, but it is not, in the writers' opinion, a totally satisfactory definition. Before we can say whether fitness for purpose has been achieved, we need to know what exactly the purpose is and how fitness in terms of factors such as performance, duration, reliability, accuracy and so on is to be measured. Moreover, to enter into a contractual undertaking to provide something on the definition of fitness for purpose alone would be unthinkable; it could be almost impossible to demonstrate at law.

In the world of business we tend to operate on the basis of contractual arrangements between parties – client and consultant or consultant and contractor, for example. These are (or should be)

7

based upon clear definitive specifications which cover what is to be done or provided, standards to be met, codes of practice to be used and so on.

The definition of quality we would therefore prefer to use is 'The complete meeting of customer/client requirements'. This enables us to embrace both the ultimate *external* customer and also the *internal* customer, namely the person(s) who receives our output within the process chain.

We are all simultaneously both providers and receivers and our ability to satisfy our external customers is directly related to how well we recognize and satisfy the needs of our internal customers.

Quality Assurance (QA)

The evolution of QA from techniques of final inspection in the 1930s, followed by quality control, mainly in the manufacturing industries, during the 1940s and 1950s, then a further extension of controls into the engineering/design phases of these industries during the 1960s, is well documented.

The realization that the greatest influence on result integrity lay in the areas of research, design and planning, and not necessarily in the subsequent manufacturing and construction phases, led to the generation of the early quality standards which to this day with the current BS/EN/ISO 9000: 1994 are not significantly changed in principle, but are more a reflection of changes in technology.

There are numerous quality standards, all of which define in almost identical terms those things which an organization should recognize if it aims to achieve quality consistently and in a cost-effective manner. What standards such as ISO 9000 tell us very clearly is that:

1. Business is a team involvement, in which everyone should play their part and know what their part is. (This indicates clear organization and responsibility information.)
2. There needs to be a policy for quality attainment set down by management and understood by all. (This indicates clear management initiative and training for all.)

3. Activities need to be carried out in ways which are as effective and as efficient as possible for the business concerned. (This indicates the need for properly thought-out procedures and instructions and training for those who are to implement them.)
4. Business scopes and technologies change and therefore any management system must be dynamic, updated and monitored for relevance to current needs and continuing adherence to its requirements by all concerned. (This indicates the need for ongoing review and audit.)

The various clauses of the quality standards (20 in BS/EN/ISO 9001) identify those things which should be considered with respect to both specific and varying scopes of activities, when developing and using a quality system (see p. 10).

The term 'quality assurance' unfortunately tends to make people think of the finished product and services, whereas it is something far greater; in fact, today's quality system is not something imposed on top of other business systems, it is *the system* of the business.

Our definition of quality assurance is: 'a structured approach to business management and control, which enhances the ability to consistently provide products and services to specification, programme and cost'.

Business objectives

Businesses are judged on performance, which is essentially a measure of achievement against objectives. Performance, however, is measured differently, depending upon the viewpoint. For example, a client will normally judge performance by the ability to receive agreed products or services to programme and price, whilst your own board may judge performance against the criteria of profitability, return on investment or capital employed, market penetration, increased turnover, customer satisfaction, and so on.

Regardless of viewpoint, performance is enhanced when there

are sound management and control disciplines and effective and efficient working methods, implemented through a trained and committed management and workforce. *That is the quality assurance approach.*

The quality system

The formal Quality System is the documented expression of management's quality policy. Many who embark upon the task of establishing a Quality System within their own organization make that task more onerous than it needs to be.

In reality the techniques which enable an organization to set up a Quality System *which is right for the organization*, whilst still satisfying the *applicable* requirements of a quality standard such as ISO 9001, are straightforward and we will discuss them in some detail in Chapter 4 of the book.

First, however, a few words of advice.

1. *Always start from the top.* Setting up a quality system is likely to be one of the most important steps an organization will ever undertake. The implications for business performance and competitive edge are very real. It is essential that top management is aware of the significance of QA and shows real commitment to establishing the system. Objectives need to be set and resources allocated. This requires top management approval. It is futile to embark on the QA route with half-hearted or 'as soon as possible' attitudes.

2. *Don't sacrifice good working practices* unnecessarily to comply with a rigid interpretation of BS/EN/ISO 9000. Understand that the standard is not a rigid set of rules which must be slavishly followed. It merely identifies a number of very important, good management and control practices, which if adopted will enhance both result quality and consistency, and improve business performance. Study it, understand it, apply its principles, but do not dismantle good, well-proven techniques purely to restructure things in accordance with the 'as written' statements of the standard, when it is totally

unnecessary to do so. Remember there is no such thing as the standard organization. Each one is different in the way it is structured, resourced, located, and so on.

3. *The system you want is the one that is right for you.* There are those in the market who offer stereotyped quality manuals, (usually word-processed documents) as the short route to QA success. Ask yourself, 'What do such people really know about *our* organization, its strengths, weaknesses, objectives, intentions, resources, system, organizational structure, and so on?' Then ask yourself, 'Would a bought-in system from such a source really be of value?' We leave you to reach your own conclusions.

4. *Involve as many people as possible in developing the system.* The important question about your QA system at the end of the day will be, 'Does it work?' It is only people that cause things to happen; they are the motivating force. It is essential that everyone therefore understands what the system is about, knows how it affects them and appreciates why it is important that they play their part.

   Participation during systems preparation and introduction greatly helps promote the feeling of contribution and understanding, and will help to ensure commitment later when the system becomes a working reality.

5. *Don't ignore training* for all levels of management and staff at each key stage during the development and introduction of the system. For example, ensure senior management has a common understanding of the QA objectives from the outset. Ensure managers in turn get the *correct* message to their staff. Ensure that people who may be involved in procedure drafting are trained in such techniques as information gathering and know what constitutes a well-thought-through and written procedure. Ensure people understand *management's interpretation* of the system (eg procedures) prior to the formal introduction of the system. Commitment to the system will not be achieved by imposition (an all too frequent approach), but by real understanding of its requirements brought about by proper top down communication and training.

6. *Don't rush for third-party certification.* First think seriously about both the benefits and implications. Although some client bodies seek evidence of certification, it is not a requirement of quality standards such as ISO 9000 that a quality system be certified. Other than probably ensuring that you *do* keep your QA activities under close review (audit), certification will not necessarily improve the workability or effectiveness of your system. Third-party certification for example, represents an initial monetary outlay, followed by repeated costs as subsequent reviews take place.

The main benefits of certification have to be seen from the marketing point of view. If having a third-party assessed quality system is definitely going to enhance your market penetration and help you to win business from clients who seek evidence of such assessment, then go for it. Do not however charge the costs against your quality department (or similar) budget. Costs should go against marketing, sales or publicity because that is where they belong.

3 Quality Assurance Standards

Background

Quality standards as used by many sectors of British industry today evolved from the defence industries. The Ministry of Defence introduced its own 05–20 series of defence standards in the late 1960s. These MOD standards were, however, only applicable to organizations contracted to the Ministry of Defence.

In 1974 the British Standards Institution (BSI) introduced a guideline (BS 5179) for organizations producing their own quality assurance documentation. Following BSI's lead, a number of organizations seeing the benefits of using quality standards to express a requirement for management control when placing enquiries and contracts, produced similar standards. These included CEGB, British Gas, and the National Coal Board. (At this time no organizations in the construction industry had developed QA standards.)

In 1979, in an attempt to rationalize the situation, BSI withdrew the guideline and introduced BS 5750 which was a three-part standard, with part 1 representing the highest level and part 3 the lowest. As a result, most organizations then withdrew their own company standards and adopted BS 5750.

Since 1979, other countries have adopted BS 5750 and modified it for their own national standard. In 1987 the International Organization for Standardization (ISO), under the chairmanship of the Canadians, prepared a series of documents, ISO 9000 to ISO 9004, ISO 9000 and ISO 9004 being guideline documents and ISO 9001 to ISO 9003 quality system specifications, using BS 5750 parts 1 to 3 as a basis for parts of the standard. After consideration of ISO 9001–9003, the committee of BSI adopted the ISO standard and reissued BS 5750 parts 1–3 as dual standards.

In 1994, in the light of user experience, the standards were reviewed and reissued. With these latest amendments and in the interests of international harmonization and trade, BS 5750 as a number was withdrawn and the new standards numbered and titled BS/EN/ISO 9000 as follows:

BS/EN/ISO 9001:1994	Quality Systems – Model for quality assurance in design, development, production, installation and servicing
BS/EN/ISO 9002:1994	Quality Systems – Specification for production, installation and servicing
BS/EN/ISO 9003:1994	Quality Systems – Specification for final inspection and test.

Table 3.1 shows the content of the three parts of the standard.

Quality standards in the construction industry

Since 1979, mainly due initially to the requirements on nuclear power stations, the construction industry has gradually adopted ISO 9000 for use by its main construction contractors, subcontractors and suppliers. It is interesting to look at the three parts of ISO 9000 and see how they are applicable to the construction industry.

Requirement	9001	9002	9003
Management responsibility	4.1	4.1	4.1*
Quality system	4.2	4.2	4.2*
Contract review	4.3	4.3	4.3
Design control	4.4	4.4	4.4
		(n/a)	(n/a)
Document and data control	4.5	4.5	4.5
Purchasing	4.6	4.6	4.6
			(n/a)
Control of customer-supplied product	4.7	4.7	4.7
Product identification and traceability	4.8	4.8	4.8*
Process control	4.9	4.9	4.9
			(n/a)
Inspection and testing	4.10	4.10	4.10*
Control of inspection, measuring and test equipment	4.11	4.11	4.11
Inspection and test status	4.12	4.12	4.12
Control of non-conforming product	4.13	4.13	4.13*
Correct and preventive action	4.14	4.14	4.14*
Handling, storage, packaging, preservation and delivery	4.15	4.15	4.15
Control of quality records	4.16	4.16	4.16*
Internal quality audits	4.17	4.17	4.17*
Training	4.18	4.18	4.18*
Servicing	4.19	4.19	4.19
			(n/a)
Statistical techniques	4.20	4.20	4.20*

* Indicates a less stringent requirement than 9001 and 9002
Source: BS/EN/ISO 9001, 9002, 9003, with the permission of BSI,
   Linford Wood, Milton Keynes, MK14 6LE

Table 3.1   Index of requirements in BS/EN/ISO 9001, 9002 and 9003

ISO 9001

As can be seen from Table 3.1 this is the most onerous of the QA standards and the only one which covers 'design'. Therefore any organization involved in conceptual design which was setting up a quality assurance system would set up its management system to meet ISO 9001.

This specification would apply to large building and civil companies carrying out their own design, to design consultants, architects and so on.

ISO 9002

This would be the standard applicable to manufacturing or installation organizations not carrying out conceptual design activities, or whose detail drawings were being approved by the client or main contractor. ISO 9002 demands process control and therefore would apply to many construction contractors where evidence of interstage inspections and tests has to be given to the client, for instance:

● construction of formwork and falsework
● placing concrete
● manufacture of precast concrete blocks
● laying damp-proof membranes
● building structures where there are hold points for inspections and checks at various stages.

ISO 9003

This is the least onerous level of the standard and is applicable to organizations setting up management systems to control their activities, and wanting to be able to demonstrate to a client the adequacy of their product or service by final inspection and/or test.

The nature of the product or service does not require a documented system to cover in-process controls but ISO 9003 could typically apply to the following product areas:

● manufacture of machined components
● manufacture of galvanized ductwork

- building of a non-load bearing brick wall
- laying of roofing tiles.

Interpretation of ISO 9000
Because ISO 9000 is a standard which can equally be applied to the manufacturing industry, the service industry and the construction industry, it can be interpreted in different ways by different sectors.

The following section looks at ISO 9001 in detail to show how it can be applied in practice to the construction industry.

A detailed analysis of BS/EN/ISO 9001: 1994

Management responsibility (4.1)
The responsibility for *establishing* and formally *defining* the quality policy and objectives is placed clearly with management (with executive responsibility). It further requires management to ensure that the policy and objectives are understood at all levels.

This means that there has to be visible commitment by management to quality, and that it is management's responsibility to see that the message is imparted and understood through proper training, explanation or other appropriate techniques.

Once managers have put the message across and satisfied themselves that all are aware of it, it is incumbent on them to ensure that the policy is being properly implemented and commitment is being maintained.

What does this mean as far as the construction industry is concerned? It means that we need to review the current policy statement to ensure that the company's quality is relevant to achieving organizational goals, and the expectations and needs of the customer. This policy must be understood by all levels in the organization.

Management can ensure that the system is being applied continually and effectively in several ways. (see 'Management review (4.1.3)' on p. 19).

Organization (4.1.2)

Responsibility and authority (4.1.2.1)   Emphasis is placed on the responsibilities and authority of those engaged in managing and performing work, resolving quality problems by analysis, solution application and verification, and in ensuring corrective measures are implemented. This section also concerns those involved in control of known non-conforming products pending disposition decisions.

In the construction industry, site organizations are set up to meet the requirements of a specific contract. These organizational structures require to be defined on organizational charts backed up by a text defining responsibilities of key people.

It may be necessary to relate the site organizational chart to the head office organizational chart if interfaces exist.

Resources (4.1.2.2)   This section of the standard requires that adequate resources are provided including the assignment of trained personnel (see 4.18) for the management, performance of work and verification activities including internal audits (see 4.17).

What is the significance of these two subsections? We are being asked to define our O & R structure but with an emphasis on persons having verification, fault analysis and corrective action monitoring roles.

Again the importance of training is stressed. Training needs to be visible and recorded. Such actions as formal teach-ins on procedures, tests of understanding and records of such training are now becoming the norm in many companies. This may mean to many an extension of training policy but it is something well worthwhile and cost effective. Some may consider education costly, ignorance is far more so.

Management representative (4.1.2.3)   The company's quality executive shall appoint a member of staff, who irrespective of other responsibilities, has defined authority for:

● ensuring that the quality system is established, implemented and maintained and for reporting on the performance of the quality system to the company's management.

There is no requirement however for this post of 'Quality Manager' to be a full-time role.

Management review (4.1.3)

This is, by definition, a management responsibility. Reviews should take place at 'appropriate intervals'. There is therefore a requirement to define both the basis and frequency of reviews, and, whilst from an organizational and responsibility point of view, it should be clear *who* decides this.

The executive responsible for the quality system has to be involved with these reviews, and ensure that the requirements of ISO 9001 and the company's stated quality policy and objectives are reviewed.

It is further required that records of such reviews are kept, which means formally logging findings, actions and outcomes (although these do not have to be made available to the purchaser representatives). The standard differentiates between independent quality audits and the management review; the latter should, however, include recognition of audit findings.

As far as the construction industry is concerned this indicates the need for procedures to define how management plans, conducts and records reviews. It requires that the reviews do consider the effectiveness of corrective and preventive action.

The reviews committee should therefore receive feedback on such items as:

- non-conformance's and corrective/preventive action effectiveness
- internal audit findings and corrective action effectiveness
- customer complaints
- conformance to organizational goals
- proposed changes to the quality policy/system.

Quality system (4.2)

There is a requirement to produce a documented system which

reflects the scope of work of the organization or project/site teams.

It should be unnecessary to create a lot of paperwork, the idea being to 'make paper work'.

The documented system of procedures instructions, method statements, checklists, standard forms and so on should be consistent with the level of control required. Procedures should *not* be produced if simple instructions will suffice, and instructions should *not* be written if a simple checklist or standard form is adequate.

The quality system will normally be documented in the form of a quality manual or manuals or in the form of a quality programme (see p. 65).

The documented system should describe how planning is to be carried out to meet quality objectives.

Methods used in the construction industry for carrying out planning are by the use of quality programmes and quality plans.

Contract review (4.3)

This is a very important clause. It is obviously beneficial to both parties in a contract that there is a clear understanding of what is to be done, how it is to be verified, and so forth. A little time spent at the beginning dotting the i's, crossing the t's and ironing out anomalies can be a beneficial investment. The situation of never having time to get things right at the outset, but always prepared to find time (usually at much greater cost) to put things right later on is all too common.

Although, at first glance, it may seem to be principally in the interests of the supplier to establish a clear brief, this is not strictly so. The standard places much greater responsibility on the client to define requirements. It is no longer a case of the supplier being expected to ask for information needed.

Contract review should be a formal activity covered by appropriate procedures. There should be a recording of review findings, along with positive mutual resolution of problems with the client and final and formal confirmation of agreed requirements. It is in all parties' interests to get things right from the outset.

Contract reviews are carried out before submission of tenders, or the acceptance of a contract or order.

Once a contract is running a contract review should be carried out on amendments to contract, to check the acceptability of the changes and transfer the information to the departments concerned.

Records of reviews are to be retained in the appropriate contract/project files.

Design control (4.4)

General (4.4.1)
This section of the standard recognizes the significance of the design function in the attainment of quality and reliability objectives and in the meeting of specified requirements. The importance of establishing the basis of a sound design cannot be overemphasized. The control of design activities to ensure proper review, planning, verification and so on is a key (if not *the* key) part of the QA system.

Design and development planning (4.4.2)
This requires that design and development activities be carried out in a planned manner and that such plans are dynamic and reflect changes as the evolving design dictates.

Key design and verification activities are to be carried out by suitably qualified persons.

Organization and technical interfaces (4.4.3)
Interface controls must be specified, particularly in terms of activities, documents and responsibilities. This is very important; interfaces between different functions are high risk areas where responsibilities are likely to become confused or break down.

Design input (4.4.4)
It is logical that the integrity of the design depends to a high degree upon the clarity and understanding of the input information. This understanding should not be left to chance, but established and agreed. Design inputs include statutory and technical

requirements which must be reviewed for adequacy. Design inputs shall take into account contract review activities (see 4.3).

Design output (4.4.5)

It is important that design outputs are clear, unambiguous and expressed in the correct units or terms corresponding to the purposes they will be used for. They should be clearly seen to meet specified input requirements fully. Design output shall be documented in terms that can be verified and validated. Design documents shall be reviewed before release. The level of review depending on factors such as safety criteria.

Design review (4.4.6)

This clause is now a mandatory part of the standard and requires that at appropriate stages of design, formal documented reviews of the design activities shall take place. Participants at these reviews shall represent all disciplines involved with the design. Records shall be maintained.

Design reviews should give the design team and the client confidence that the technical aspects of the design have been thoroughly analysed, before carrying out further design or moving into the construction phase.

Design verification (4.4.7)

Verifying that design outputs do reflect input is important to ensure that subsequent work does not proceed on inaccurate information.

Verification should be carried out by suitably qualified designated persons. Typical verification techniques include checking, carrying out alternative calculations and proof by test.

Design validation (4.4.8)

Design validation has to be performed to insure that the construction conforms to the user need and/or requirements.

Techniques such as testing and commissioning can be used to validate the design.

Design changes (4.4.9)
It is important that changes to design be dealt with as thoroughly as the original design.

How is this relevant to the construction industry? Designs may be supplied by the client, prepared by an independent design contractor or by the design contractors for their own site organization. It is important to have control over the design and at the interfaces with design, for instance, planning, resident engineer, quantity surveyor. *Quality cannot be constructed or manufactured into a product or installation; it must be designed in.*

Document and data control (4.5)
This section of the standard is applicable to all departments in the organization involved with the quality function. Personnel must know what they are being asked to do, and therefore up-to-date information must be available.

The requirement relates not only to drawings and specifications, but also to procedures, instructions and so on that describe how activities are to be performed.

One important aspect of any document and data control system is to have up-to-date and controlled distribution lists for quality related documents. Thus, when revisions are made, the recipients of the original issue can receive the amended version. This control should also extend to the removal of obsolete documents from the workplace, and either destroying them or marking them in such a manner as to prevent inadvertent use.

Another important point with any documentation and data control system is to have nominated personnel responsible for amending documents and data. For example, it should not be possible for a method statement on site to be changed by the resident engineer or the clerk of works, but only by (or with the knowledge and formal approval of) the originating department.

The standard appreciates that many organizations control data on computer systems in addition to hard copy documentation. It is equally important that this data is controlled.

Typical aspects which differ from document control are:

- verification and validation of computer software
- virus checking of disks before use
- passwords to prevent corruption of data on the computer database
- control of drawings produced on a Computer Aided Design (CAD) system.

When audits are carried out to check the effectiveness of systems in the various departments, the control of documents and data is one of the more important aspects. Documentation is what the auditor will be examining to establish whether or not personnel are using laid down systems. If, for example, someone did not ensure that obsolete drawings were removed, then it could be possible for scrap to be produced.

Purchasing (4.6)

The acquiring of correct products requires the clear specification of product requirements and a sound knowledge of the potential supplier's ability to provide products of a certain quality.

Ensuring that purchased goods and services conform to the requirements specified by the order is not solely a job for incoming inspection. There is a requirement to ensure that orders are placed with suitable suppliers/contractors capable of meeting the specified requirements fully; also that purchasing documents themselves are clear and definitive before orders are released.

Companies may be selected on the basis of good products/services previously provided. Where, however, unusual or new services are required from unknown suppliers, it is essential to have some other criteria for assessing their competence. One method of doing this (especially when placing a contract incorporating a QA requirement such as some part of ISO 9000) is to do an assessment of the quality system.

Such assessments should be carried out in a controlled manner, preferably against standard checklists. By assessing against a standard, that is a recognized quality criterion, it becomes possible to compile a realistic list of approved suppliers/contractors.

An assessment might reveal that if work were placed with a certain contractor, extra supportive effort would be necessary to compensate for weaknesses in the system. The resulting 'on-cost' of this support should be recognized at the pre-order/bid assessment stage. Failure to do this could lead to an adverse 'closed order variance'.

Records of supplier capability and performance should be established and maintained. Once material arrives at the purchaser's premises its control becomes subject to appropriate internal procedures. These typically include:

- approval of quality plans
- checking of documentation such as material and test certificates and so on.

Where specified in the contract the customer's representative has to be afforded the right to visit the subcontractor's premises, although this will not absolve the supplier of his responsibility to provide acceptable products.

Remember that the degrees of control should be commensurate with the specified requirements for the product or service in question. For example, do not specify a standard such as ISO 9002 within a product specification if its quality can be assured by final inspection and test only.

Control of customer supplied product (4.7)

Supplied products are usually the responsibility of the receiver once they have been accepted. It is, therefore, essential to have controls which ensure that any product found unacceptable is returned, and that all products accepted are controlled in accordance with laid down procedures and contractual requirements.

Once the products have been accepted, for instance by goods inward inspection or the resident engineer, they should be controlled in some way to prevent use on work other than that for which they were obtained. Products must be suitably identified and stored in a manner to avoid damage and deterioration.

Sites are notorious for equipment being lost, damaged or issued against some other contract. It is, therefore, very important that when 'free issue' material becomes your responsibility, suitable controls are established and implemented.

Where the client supplies mechanical plant items, such as travelling cranes, for use in the permanent works, they are usually accepted on the basis of a visual examination and no functional test is carried out unless by contractual requirement.

It has to be borne in mind that the customer-supplied product is not necessarily related to items delivered on site. It could be computer software or hardware used for design purposes.

Product identification and traceability (4.8)

This clause begins with the words 'where appropriate'. This may seem a little subjective, but there are situations where traceability will be both essential and obvious. For example:

● when the specification or contract requires it
● when purchaser supplied material has to be used for specific applications
● when special materials (such as special chemical compositions) have to be released or used only for defined applications.

The importance of this section is that there are procedures which can ensure that traceability requirements can be identified and catered for throughout or during specific stages of work programmes.

In many cases it is the suppliers who must take steps to ensure that they operate their own systems to ensure that the product is identified where appropriate, and is backed up by the necessary documentation.

It is obviously inappropriate and very costly to have full traceability of all material used on site. The client must therefore be made aware of the impracticality of requesting full documentation and identification of all materials used unless really necessary. A concrete nuclear pressure vessel could be a case for full identification and traceability of materials.

Process control (4.9)

This section relates to the planning and controlling of the processes within the organization's scope of work.

These controls include:

- documented procedures
- use of suitable equipment
- compliance with standards and regulations
- monitoring and control of process parameters
- criteria for workmanship
- suitable maintenance of equipment which affects the process.

Process control must be employed on all disciplines, for example:

- planning
- estimating
- design
- manufacture
- installation
- testing and commissioning.

Particular emphasis should be given to processes affecting product quality which cannot easily be measured, or entail special skills or where results cannot be fully verified by subsequent inspection or test. In these circumstances, aspects which need particular attention include:

- the integrity of measuring devices
- the training and competence of operators
- control of environmental conditions
- certification records.

What does this mean to the construction industry? Workmanship standards have been established over many years in the construction industry, although not necessarily documented as required by a quality assurance regime.

Some contractors have had their materials controlled under the BSI kitemark scheme or the Agrément Board Certification, but in other areas there has been little evidence of formal control.

Where special processes demand more than a simple instruction and method statement (that is, where there are interstage inspections and/or verifications by the client), it may be necessary to prepare a quality plan for submission to the client for approval before production or installation commences. Typical examples of situations in which quality plans may be requested by the client are:

- placement of reinforcement
- formwork and falsework
- laying of waterproof membranes
- manufacture of precast concrete blocks.

Inspection and testing (4.10)

General (4.10.1)
All parts of this clause emphasize the requirement for inspecting against a defined base such as a quality plan (which may in turn reference drawings, inspection checklists and so on as criteria) or documented procedures. In other words the emphasis is now not only on *what* is to be inspected, but very much on *how*.

Receiving inspection and testing (4.10.2)
This clause deals with the requirement to control incoming product or materials. Procedures or plans should define the level of verification required to meet the specified requirements. This may vary from a visual, examination of product and documents to a full inspection and/or test.

In determining the level of receipt inspection consideration must be given to the amount of control exercised by the subcontractor.

In-process inspection and testing (4.10.3)
This clause is significant, requiring product verification by direct inspection and test and by monitoring the process used to produce the product. This means planned inspection, test and monitoring and not informal subjective techniques. It is a requirement that product shall only be released (unless under controlled conditions) when the appropriate inspections and test have been completed.

Procedures and instruction should require this, and identify controls that must apply if exceptions are necessary.

Final inspection and test (4.10.4)
This clause makes it clear that a mandatory requirement for final inspection and testing is that all previously planned inspections and tests have been completed and that the evidence and results of these are satisfactory.

Construction work shall not be continued until all planned/ required activities are known to have been satisfactorily completed and supporting authorized evidence is available. (In many cases this would be a signed-off manufacturing or installation quality plan.)

Inspection and test records (4.10.5)
This emphasizes the need for the generating and maintenance of appropriate records. (Again in many cases this would be satisfied by final signed-off quality plans and their supporting documentation.)

Records should clearly show whether items have passed or failed. If the product has failed it should be evident that the procedure for control of non-conformances has been applied.

The very nature of the construction industry makes interstage inspections very important because it is often impossible to verify the work at the final inspection stage.

In the local authorities it has been traditional for the building inspector to check work at various stages against the building regulations. Likewise on other building sites inspections and tests need to be made, such as:

- checking the cleanliness and stability of shuttering systems
- checking brickwork at stages with a theodolite
- checking reinforcement bar prior to placing concrete.

Evidence of these inspections and tests can be by completion of inspection checklists; signing off the quality plan, preparation of a test certificate, and so on. Construction staff can usually inspect and document the facts themselves, although there must be some supervisory level, verifying on a percentage basis and signing off the relevant documentation.

Control of inspection, measuring and test equipment (4.11)
All measuring equipment used for inspection and testing must be in a known state of calibration. To ensure this, the following are required:

1. Check measuring and test equipment at regular intervals determined by its usage.
2. Identify each piece of checking equipment for record and control purposes.
3. Record the date of calibration and date of next due calibration for each piece of equipment.
4. Establish a recall system which ensures that items of equipment are taken out of service and calibrated at the specified intervals.
5. Use correct procedures for checking equipment.
6. Ensure that the equipment used for calibrating is itself within the calibration system, for instance, checked by an approved laboratory or test house.
7. Relate all calibration facilities back to national standards or the basis for calibration.

What does this mean to the construction industry? Inspection equipment such as levels, theodolites, batcher plant instruments, cube testing equipment and tapes must be in a known state of calibration.

Certain instruments and equipment must be tested regularly to give confidence that the inspected or tested items meet the requirements. For certain equipments the calibration, as part of the planned maintenance programme, may prove cost effective; for example, a batcher plant that is mixing incorrect quantities of materials can prove very costly, compared with regular maintenance and checks on the plant.

Inspection and test status (4.12)
If required, it should be possible at all stages of construction to be

able to establish precisely the inspection status of material, components or assemblies. There should be a way of knowing at all times whether the product or installation has:

- not been inspected
- been inspected and approved
- been inspected and rejected
- is awaiting concessionary action.

Therefore approved markings, either on the product or the assembly, and/or inspection records should be evident to distinguish between inspected and uninspected products. Procedures should state clearly what the method of identification is and how it is to be marked on the product.

As far as the construction industry is concerned this requirement of the standard is straightforward in some areas, that is where a product is being produced; for example, reinforcing steel and precast concrete blocks. However, it is not easy to identify by labels or tags the inspection status of a pour of concrete, a large item or a structure. Here you have to rely on documentary evidence to identify, for instance, the area of the structure where a pour of concrete was made.

Control of non-conforming product (4.13)

Defective products must not be allowed to be used. Even the best controlled systems produce suspect or unsuitable material; the system should be such that this is identified and controlled. Any defective work whether unsuitable, awaiting rework or awaiting concession, should be set aside and identified to prevent inadvertent use. The system used should ensure that information about such products is fed back to the appropriate parties so that action can be taken to prevent recurrence.

Written authorization should be given before using any product which does not conform to specification. In the case of products purchased from subcontractors that are non-conforming, the purchaser should agree any concessions in writing before permitting their use. Repaired work should be reinspected.

How does this apply to the construction industry? If material on site is found to be rejected, suspect or without documentary evidence of its suitability, measures must be taken to prevent its use in the permanent works. The permanent works may also be found to be outside the specification, and then measures should be taken to prevent further work until the problem is corrected.

In cases where a large batch of concrete has been mixed and then found to be outside the design mix, it may be possible (on agreement of the authority) to use this material elsewhere. The system should be such that a quick decision can be reached, so that material such as concrete does not become unusable.

Corrective and preventive action (4.14)

To prevent the recurrence of faults, prompt and effective action must be taken to identify and correct the causes. Defective work can have many sources – incorrect or lack of work instructions (including drawings), faulty materials, tools or methods, poor design, human error and so on.

Responsibilities should be clearly defined so that when anything goes wrong action can be taken at once through the proper channels. It is not enough to put aside the defective work and correct it. The cause of the defect must be found and corrected, otherwise the same fires are fought repeatedly.

Problems may arise with products or equipment provided by suppliers and subcontractors. The system should provide for departments responsible for checking such goods and services to initiate corrective action.

Documentary evidence for all corrective action and the reason for it should be retained.

Corrective action may involve modifying designs, specifications or working methods, or enforcing conformity to instructions which can require many departments to initiate and implement changes. For example:

- The design office would deal with concession and design change requests, changes to specifications, drawings and so on.

- The quality assurance department would amend quality related procedures and instructions.
- All managers would ensure compliance with laid down procedures, instructions, specifications and so on.

What does this mean to the construction industry? It is important that the right authority makes the corrective action decisions. In the construction industry, problems with drawings or specifications should be fed back to the design authority. If there are problems with materials, then the suppliers need to be involved to ensure steps are taken to prevent the same fault occurring in the future.

Corrective action systems that are in place on site must be such that a quick decision can be reached. The responsible person (such as the clerk of works) often has to make an urgent decision as to what corrective action to take, and then complete the necessary paperwork.

Where an urgent decision is required from the design authority to prevent construction work coming to a halt the fax system may be used.

Preventive action extends to analysing results of work operations, concessions, audit reports, quality records and customer complaints to detect and eliminate potential causes of nonconformances. Relevant information should be submitted to management review meetings (see 4.1.3).

Handling, storage, packaging, preservation and delivery (4.15)

From the receipt of material through to the final despatch of the finished parts (and, in some cases, installation) controls should ensure proper identification, inspection, handling, preservation and storage to prevent damage or deterioration.

In the case of some special materials, such as stainless steel, it may be necessary to segregate it to avoid contamination from other materials. In other instances, it may not be possible to use standard procedures for identifying material; for example, hard stamping may not be permitted and special marker pens may have to be used on exotic materials. Any special contractual requirements should be transmitted to the parties concerned.

Any system for the protection and preservation of product quality must control access to areas where material is being held pending use or shipment. Suitable procedures for receipt, recording, storage and despatch of material help ensure effective control.

In the construction industry the waste of materials due to poor storage and handling on site has been very costly for both contractors and clients in the past.

Money can be saved by adequate forward planning concerning:

- what materials are going to be required on site
- when materials are required
- what storage conditions are required
- what manpower resources are required to handle, store, pack and deliver material.

Controls are required over the security of material on site to avoid loss and damage, and to avoid the issue of material against another contract.

Control of quality records (4.16)

Records are necessary in all departments to provide objective evidence that systems are working effectively. They should be controlled in such a manner that it is possible readily to retrieve and analyse any documents/data in the system. For this reason, and for the benefit of those working in the department, records should be indexed and stored in a suitable manner. Records should also be made available to the customer's representative where agreed contractually.

The following are typical of the records required in the construction industry:

- Agent/contracts office:
  - i   tenders and contracts
  - ii   minutes of meetings internally and with customers when contractual decisions affecting quality are made

     iii   subcontractor's records if part of contractual requirements

- Design office:
  - i   record of latest issue of all drawings
  - ii   concession applications and responses
  - iii   design change requests and responses
  - iv record of standards available
  - v   feedback from operating plant
  - vi   records of technical queries
  - vii site visit report
  - viii approvals of contractor's shop drawings
  - ix   minutes of contract review meetings
- Inspection and test:
  - i   inspection reports
  - ii   calibration records of all measuring and test equipment
  - iii   test certificates
  - iv completed concession forms
  - v   completed checklist and test schedules
- Quality assurance:
  - i   audit reports
  - ii   material certificates
  - iii   welder qualification records
  - iv training records
  - v   list of approved contractors
  - vi   quality plans.

Internal quality audits (4.17)

Planned programmes of quality audits need to be carried out to verify the ongoing adequacy and effectiveness of the quality system. Audits and their follow-up actions are to be conducted in accordance with formal procedures.

Results of audits are to be notified to the appropriate responsible management personnel who shall take effective corrective actions in a timely manner.

Audits are an independent activity and must be carried out by individuals independent of the area being audited, who must be trained in auditing techniques.

In the construction industry audits must be carried out to an audit schedule. This schedule may need to be agreed with the client if the audits are for different phases of work over the life of a contract/project.

It is not necessary for the QA representative to carry out all audits: he may draw upon other staff (maybe from head office or other parts of the construction team) to carry out specific audits.

Audits are a means of monitoring the management system, and it is important that any non-compliances are corrected in a timely manner by those responsible for the area concerned.

Training (4.18)

All personnel carrying out quality related functions should have received the training necessary to ensure that they can perform their tasks. Particular attention should be given to the training of new personnel.

It is up to the manager to identify the training needs of his staff and see that the appropriate skills and experience have been obtained before designating tasks to them. In certain instances, levels of competence can be demonstrated by examination and certification either in-house or by a recognized outside body.

For some jobs the validity of the certification has to be demonstrated to the customer; for example a welder being approved to BS/EN 287 part 1 1992.

What does this mean to the construction industry? Training does not just relate to people carrying out tests where a certificate of competency can be obtained, such as welders and inspectors using specialist equipment. A site manager should consider the training required by other staff on the construction site, such as PVC waterbar welders, operators of batcher plant and constructors of formwork and falsework. The training requirements should be identified by the site manager and take into account the Health and Safety at Work Act.

Another important area where training is required is in the use of site procedure, instructions and method statements – the Quality Assurance System on site – to ensure that everyone is singing from the same hymn sheet.

Servicing (4.19)

Servicing should be carried out in accordance with the necessary procedures and instructions – service manuals, servicing schedules and checklists. Records should be available to show that servicing has taken place to the contracted requirements.

Special service books may be required. Any measuring and test equipment should be in a known state of calibration.

This clause is not generally applicable in the construction industry unless the contractor is providing ongoing servicing after the warranty period.

Statistical techniques (4.20)

Statistical techniques are useful in many circumstances, where the product or the process needs controlling.

Due consideration should be given to the use of statistical techniques for the verification of process or product characteristics; for example, sample plans for assessing the integrity of large batches of similar components.

What does this mean to the construction industry? Statistical techniques have their place in the construction industry, for instance:

- sampling of precast concrete blocks
- sampling by cube and slump test
- testing of rebar.

Sampling checks not only the consistency of the product but the capability of the equipment or plant producing the product – batcher plant, moulds for precast concrete block and so on.

4 Developing and Implementing a QA System

Establishing awareness

We have already said that QA must be introduced with the understanding, enthusiasm and commitment of top management, starting with the chief executive. There can really be no other way.

Most companies looking at QA for the first time will be doing so because of external pressures and influences. They may be receiving requests at the tender stage to provide evidence that they are operating in accordance with a QA system. Or they may be aware of others, probably including competitors, who are moving towards the adoption of QA, and they feel that if they don't do likewise they may lose their ability to compete. This is almost certain to be the case.

The discerning executive will be aware that pressure for QA is building up because those who are placing contracts for work are realizing more and more that the prospects of getting things done properly, to programme and costs, are greater when the contractor or supplier operates sound management and control systems to a recognized quality standard such as ISO 9000.

The executive may ask 'Why should QA help a company consistently to provide the right goods and services to programme and costs? The answer is that QA is about:

- the proper evaluation of requirements and their agreement between the parties concerned at the outset of a work programme
- the planning for both the quality of result and the quality of performance in achieving that result
- the identification and timely utilization of resources
- doing things in ways which have been properly considered and are the most effective and efficient (and therefore the most cost effective) for the company
- always ensuring adequate (but minimum) controls are applied at all stages of work programmes to provide ongoing confidence in both performance and results
- treating everyone who receives the output of another as a customer who needs to be satisfied
- providing evidence at the end of the job that those things which were to be done, were indeed done and done properly
- harnessing the understanding and commitment of management and workforce to act together for the benefit of the total business performance.

In other words, the main incentive to adopt QA is the real improvement of business performance that it can bring about.

Once the recognition of QA benefit exists at the most senior management level, then the prospect of successfully introducing a system becomes a reality, because it will be realized that:

- the subject is too important to treat half-heartedly
- priorities will need to be given to the system development tasks
- resources will need to be committed to achieve the tasks.

Widening the awareness

The next step is to gain the understanding and support of senior and middle management concerning the QA exercise.

As well as being given the senior executive's own reasons for going ahead with QA, it is strongly recommended that, at this stage, senior and middle management personnel should be provided with some basic QA appreciation training. A short in-house course is the most cost-effective way of providing such training, preceded by a discussion with a reliable training consultant so that the training meets the specific needs of the industry or company concerned.

A suitable one-day event to provide the level of understanding needed at this stage would address such aspects as:

● what QA is
● how it helps the achievement of business objectives
● the significance and applicability of quality standards such as ISO 9000
● preparing a system with emphasis on management role
● winning company-wide commitment
● cost-effective use of QA
● review and audit techniques
● ongoing improvement.

Such training is important because there needs to be a common understanding so that managers can, in turn, communicate a consistent message to their own staff and answer confidently the questions that will inevitably arise.

Quality manuals

The usual way of presenting the Quality System is in the form of a Quality Manual or Manuals. In the smaller company with fewer departments and interfaces a single manual will usually be adequate, and its contents will normally include:

Part 1

- contents list
- company profile
- amendments record
- policy statement
- description of how the relevant criteria of the applicable quality standard (eg ISO 9000) or part thereof are addressed by the company quality system
- organization structure figures (ie company and departmental) and supporting descriptive texts
- index of quality procedures

Part 2

- procedures
- supporting instructions
- samples of standard documents correctly completed (eg concession application, technical query).

Such a manual structure enables the document to be used to maximum effect. A quality manual normally serves two purposes: first, it provides evidence to people such as clients, of the company's QA capability and awareness; second, it provides a statement of management and control requirements for those within the company.

With the above structure, Part 1 of the document can be sent to those external to the company, without divulging detailed procedure information which should be regarded as commercially confidential and a unique statement of the way the company operates. If procedures do ever have to be sent to people outside the company it should only be under clearly defined conditions of confidentiality.

In the larger organization it is common to have a multitier hierarchy of manuals, such as:

A policy manual.   As the name implies, this is a statement of quality policy and compliance, with little or no value as a control document for actual work; it is essentially a marketing document. Typical inclusions in such a manual will be:

- index
- company profile
- amendment record
- policy statement (company)
- statement of compliance to the chosen quality standard
- company organization structure figures and descriptive text
- an index of subordinate (departmental) manuals.

Departmental/discipline manuals.   Typical inclusions in each will be:

- index
- amendment record
- policy statement (departmental or discipline with reference to overall company policy)
- departmental/discipline organization chart and texts
- index of procedures and procedures proper: *corporate* – general across the company, and *specific* – the department or discipline in question
- supporting instructions
- specimens of correctly completed standard documents.

Sometimes, supporting instructions and standard forms are actually compiled into further submanuals, giving a configuration as follows.

POLICY
MANUAL
↓
DISCIPLINE/
DEPARTMENTAL
MANUALS
↓
SUPPORTING
INSTRUCTIONS
↓
STANDARD
DOCUMENTS

The structure adopted is entirely a matter of company prefer-ence. Let us now look at some of the items included in the documented quality system.

Company profile (company or policy manuals only)

This should be a brief (ideally about one page) description giving an outline of such things as:

- age of company and development
- whether part of a larger group or not
- scope of business activities, eg home and/or overseas
- nature of business, eg design, construction, manufacture
- annual turnover and staffing levels
- reference to major diversifications and achievements, such as large projects, product ranges, and so on.

The purpose of such profiles is basically to give the reader an impression of the company, its stability, substance and cap-ability.

Amendment history sheet

The purpose of this is to enable anyone wishing to read the quality manual (company or departmental) to establish the current status of documents contained within it. Quality systems are dynamic and changes are regularly made to such things as organization charts, procedures and texts, as needs dictate. The amendment sheet is an accurate record of such changes and will normally provide information under a few simple headings as shown in Figure 4.1

Varying degrees of detail may be entered onto such sheets. In some companies, detail changes to individual pages of pro-cedures may be included (see Change ref 2 in Figure 4.1). In others, reference would only be made if the whole procedure was reissued. Alternatively, detail changes may be incorporated

Change ref	Details of change	Made by	Date	Previous issue	New issue status
1	Organization chart Fig 2, now changed to include Planning Department	J Jones	2/8/95	1	2
2	Para 3 of Page 2 (Issue 4) of Procedure DD14 changed to include responsibility details	B Smith	2/8/95	4	5

Figure 4.1 Example of an amendment history sheet

on individual amendment sheets attached to each specific procedure. The way adopted should be that which is going to be the easiest for you to operate.

The important thing is to provide a clear, accurate and auditable trial to current information, so that everyone from user to independent assessor knows they are working to a common and proper basis.

Policy statement

This should reflect the organization's awareness of both market and internal expectations and its policy concerning such matters as: satisfying its customers, developing its workforce, market leadership aspirations, recognition of health and safety and environmental objectives, the significance of adherence to the quality system, etc.

The statement should be made over the signature of the chief executive. A typical statement may read as follows:

'It is the policy of the Company to

● become and remain a leading organization in the construction industry

- provide its clients with a courteous, effective and efficient service
- design and build structures such as to fully reflect the health and safety needs of those involved in the construction, use and subsequent maintenance of same
- give due consideration to environmental issues at every stage of our work processes.

In order to achieve these aims and do so in a manner which enables the Company to achieve also its planned financial and growth objectives, the Company operates within the framework of a quality management system compatible with ISO 9001–1994.

This system is expressed through this Quality Manual and its supporting departmental manuals and procedures. Compliance by all staff with the requirements of the system is mandatory.

Signed————————————————— Date —————————'
Chief Executive

Any submanuals for departments, such as design office, planning, quantity surveying, etc should (unless simply procedure manuals) also include a policy statement. This however can be relatively simple and read something like:

'It is the policy of the Company to carry out its business activities in accordance with a quality management system, which complies with the relevant requirements of ISO 9001–1994. This Manual describes how this department manages its activities in order to comply with the Company policy and adherence to the procedures, instructions and other documents contained herein is mandatory for staff within the department,

Signed————————————————— Date —————————'
Head of Dept

Statements of compliance to the chosen quality standard

This section describes in general terms how the company's system equates to the *relevant* parts of the quality standard with which it is intended to comply, for example, ISO 9001, 9002 or 9003.

It does not need to be a lengthy section; a paragraph or two for each main subject heading in the standard will usually be adequate.

It should be remembered that a company's quality system should be written around the actual scope of work carried out. In some cases not all subject headings of the standard will apply. For example, in an organization doing nothing but design work, it is possible that no more than 14 of the 20 subject headings of the standard ISO 9001 would be relevant.

The actual number may increase, however, if the design office does research and development work and is therefore involved in test supervision. If so, sections on the calibration of test equipment, test status and so on may well apply. In other design offices such subjects would not.

The writing of this part of the manual is useful for two reasons. First, it makes the company think about what it really does and how the various sections of the quality standard apply, so that important controls which might otherwise have been overlooked are picked up. Second, it provides a clear picture of the scope of your quality system, for the benefit of others reading the manual.

A typical statement in this section of the manual concerning the subject of management responsibility (ISO 9001 4.1) may read:

'The objectives of the business and its commitment to quality are incorporated in a Policy Statement issued by the Chief Executive.

The realization of the objectives is achieved through the application of a formal quality management system which defines the organizational structure of the business, the relationships and responsibilities of staff, the procedures whereby processes are effectively and efficiently carried out, the arrangement for monitoring and reviewing the satisfactory ongoing application of the system and for correcting any problems identified.

Review techniques include those of internal audit, also periodic executive management review.

The system is maintained by a dedicated Quality Manager who has a direct reporting link to higher management concerning quality matters. All staff are "trained" in the application of the system and records of such training are maintained.'

Organization and responsibility information

Although companies still have the typical management hierarchy structures, it should be recognized that the role of the manager is gradually changing as TQM philosophy becomes more widely appreciated.

Whereas the manager has traditionally been one who gives instructions and guidance, the TQM role devolves responsibility to a much greater degree to the persons actually doing the job, with the manager acting more as a facilitator. Although this will be hard to accept and may be difficult to adjust to by many managers, results have shown this to be a better approach to business success.

Change toward the TQM approach will inevitably continue and those who do not follow this route could well find themselves at a disadvantage in the competitive market in the not too distant future.

The TQM philosophy of creating a culture and an environment for ongoing improvement through people participation does not however obviate the need for a quality management system. In fact the system provides both a good springboard from which to launch improvement initiatives and a vehicle for recording and communicating any improvements resulting.

A good quality system is a fundamental for TQM.

Within any organization there is a need to know precisely at any one moment who is responsible to whom and for what. The roles of the various functions within an organization and of those

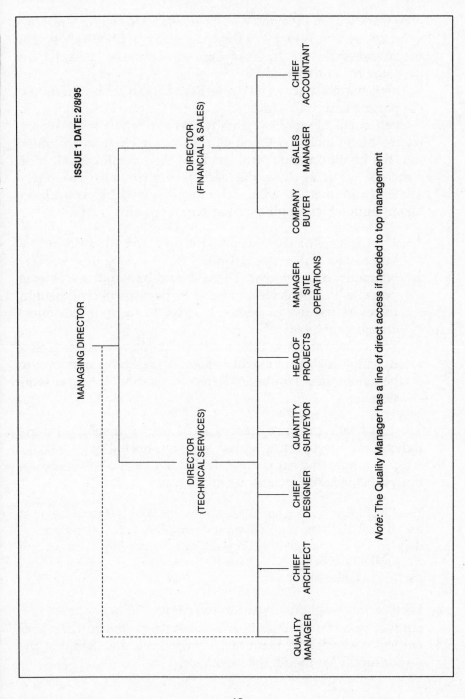

ISSUE 1 DATE: 2/8/95

MANAGING DIRECTOR

DIRECTOR
(TECHNICAL SERVICES)

DIRECTOR
(FINANCIAL & SALES)

QUALITY MANAGER

CHIEF ARCHITECT

CHIEF DESIGNER

QUANTITY SURVEYOR

HEAD OF PROJECTS

MANAGER SITE OPERATIONS

COMPANY BUYER

SALES MANAGER

CHIEF ACCOUNTANT

Note: The Quality Manager has a line of direct access if needed to top management

Figure 4.2   Example of a company organization chart

49

who work within them must be clearly defined. The essence of a good QA system is that it is always possible to identify who has the responsibility for the carrying out of specific tasks, be it a department or an individual.

This is normally achieved by the use of organization charts and supporting descriptive texts.

Even small companies usually have a number of departments/disciplines, and the first organization chart incorporated into the quality manual will often be the overall organization structure, showing how the various departments/disciplines relate to and interface with each other. Figure 4.2 shows a basic layout from which a number of important points emerge:

1. The Quality Manager has the freedom and authority necessary to do a truly impartial job.
2. No names are included. These should be avoided wherever possible. Positions tend to be more permanent than people, therefore the use of names is likely to increase document change problems.
3. The chart carries a title, an issue status and date. It is a dynamic document to which changes will sometimes have to be made; these can be indicated by altering date and issue number.

The text relating to such charts can be brief, but should clearly indicate lines of reporting and scope of responsibility.

For example, the supporting descriptive text to the organization chart for the chief designer might read:

The Chief Designer is responsible to the Director (Technical Services) for all civil and structural design and specified technical support. A detailed description of the Design Department's organization and responsibility structure is contained in Section 1 of the current issue of the Design Office Manual.

Each of the functions illustrated in Figure 4.2 will almost certainly require its own organization chart supported by a descriptive text covering the main tasks carried out and defining the responsibility for their implementation.

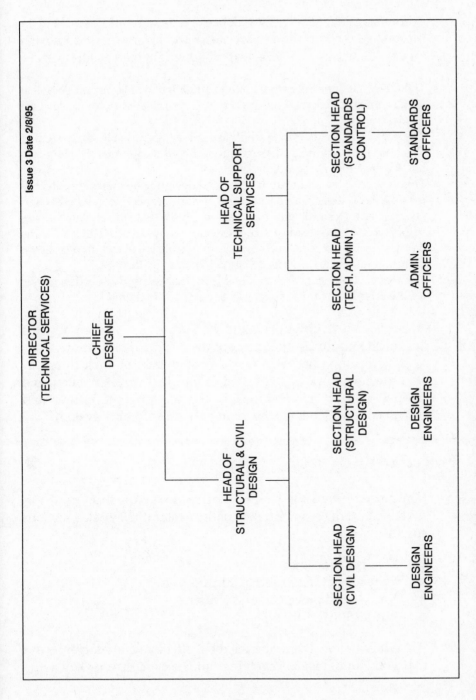

Figure 4.3 Example of a design department organization chart

51

For example, the Design Office Manual would be likely to present an organization chart along the lines of that shown in Figure 4.3, with supporting text which might read as follows:

The Chief Designer is responsible to the Director (Technical Services) for all civil and structural design activities and related technical support services.
The Head of Structural & Civil Design is responsible to the Chief Designer for carrying out structural and civil design tasks assigned to him by the Chief Designer.
This is accomplished through dedicated design teams with responsibilities for civil design and structural design respectively. The Section Head (Civil Design) and the Section Head (Structural Design) are responsible to the Head of Structural and Civil Design (HSCD) for the day-to-day supervision and control of their respective design teams in carrying out design tasks as delegated to them by the HSCD.
The Design Engineers are responsible to their respective Section Heads for the carrying out of design tasks as assigned to them.

A similar descriptive breakdown would be written for the Technical Support Services part of the design organization.

Elaboration is not necessary as the details of such things as how design work is delegated, controlled, implemented and verified will all be contained in the working procedures and instructions which form the main part of the quality system.

Procedures

Many people find the task of writing procedures the most formidable part of setting up the quality system. They ask questions such as:

- What is a procedure?
- What subjects do I need to address?
- How should procedures be written?
- Who authorizes them?

Procedures should describe the most effective and efficient ways that you can devise for carrying out the processes or key activities necessary to fulfil your customers and your own requirements. They describe the 'what', 'when', 'how', 'where' and 'who' of things and as such should be written in as 'reader friendly' manner as possible.

Adequate time must be allowed for the drafting and review of these key documents. They form the heart of the quality system and it should be remembered that what you write today may well determine the way things are done for a long while to come.

Some would advocate that you merely need to write down what you do, then do what you have written. Such an approach however is hardly likely to bring about the benefits that a QA approach should produce.

Instead write down what you do, then ask yourself:

- Why do we do things this way?
- Are all the steps really necessary?
- Is there a better, more efficient way we can do things in light of our current technology / resources?

In other words stop, think, clarify, qualify and simplify. Only when you have done this produce your draft procedure(s).

When identifying the subjects for procedures you need to go beyond merely addressing your current processes. You should also refer to the relevant part of the ISO 9000 standard that you are aiming to satisfy.

This will highlight other subjects for which you will need to write procedures such as 'management review', 'control of the manuals and procedures themselves', interval audits and others.

It is by considering what you actually do in parallel with what the standard seeks, that you ensure the adequacy and detailed content of your procedural system.

Procedure formats

There are many different formats used for procedure presentation, varying between those where all responsibilities are included within the written text, to those where responsibilities are expressed in a separate column similar to the minutes of a meeting. A further presentation is one based upon a flow diagram, with an elaboration of responsibilities adjacent to each action description.

What matters is that you choose the one that you believe is best suited to your individual company/department needs.

Regardless of presentation however, there are a number of things that should be remembered:

1. Adopt common formats for the 'front' and 'continuation' pages of your procedures.
2. Ensure that as a minimum all formats contain:
    ● procedure titles
    ● unique reference number
    ● identity of issuing department
    ● specific page number and total number of pages in the set (for example, p. 2 of 4)
    ● document/page issue status
    ● date status.
3. The front page (or as some companies prefer, a procedure cover sheet) should carry:
    ● a signature of authorization
    ● a signature indicating acceptance by the company's quality assurance authority (signifying that the procedure is compatible with QA requirements in terms of layout, style, ISO 9000 requirements and for correct interfacing with other procedures).
4. Use a consistent style for presenting the information, for example under a number of key headings as follows:
    ● **Introduction** or **Purpose** which describes what the procedure is about
    ● **Scope** which defines the parameters of its application
    ● **Procedure** which describes in logical numbered paragraph sequence and simple language, the discrete action steps which constitute the procedure.

Additional headings, such as **Reference documents** or **Records**, are also included by some organizations. It is however a matter of individual choice.

Figure 4.4 shows a typical procedure layout, including a few paragraphs of text.

STRUCTURAL DESIGN OFFICE						
TITLE	DESIGN CALCULATIONS PROCEDURE					P1 OF 3
REF NO DP6	ISSUE	ORIGINAL	2	3		
	DATE	5/3/95	6/5/95	10/7/95		
AUTHORIZED		*S Smith* *C.H. Designer*	*S Smith* *C.H. Designer*	*S. Smith* *C.H. Designer*		
Q.A. ACCEPTED		*J Cool* *Q A Manager*	*J Cool* *Q A Manager*	*B Jones* *Q A Engineer*		

1   INTRODUCTION

This procedure describes the way in which manual (ie non-computer) calculations carried out in the Structural Design Office are prepared, verified and approved.

2   SCOPE

This procedure shall apply to all calculations which have a direct bearing upon the integrity of a finished design, or otherwise as requested by the Chief Designer.

3   PROCEDURE

3.1   All calculations shall be presented neatly in black ink on official calculation sheets Form No D625.

3.2   The design base information (eg Code of Practice) used shall be clearly referenced in the right hand margin of the calculation sheet.

3.3   All pages of a calculation set shall include both the individual number of the page and the total in the set.

3.4   Etc.

Figure 4.4   A procedure layout example

Procedure drafting should, wherever possible, be done by those actually involved in the task concerned.

Initial training in procedure writing technique is recommended. It helps ensure a consistency in the drafts thus produced and eases the task of the editing authority. This is normally the person coordinating the QA system development exercise, supported by the heads of the various disciplines involved.

Procedures v Instructions

Most quality standards generalize when they refer to procedures and instructions. As a result there is considerable confusion as to what constitutes one or the other.

We define these terms as follows:

The Procedure is a higher tier document than the Instruction and it defines not only what has to be done, but also *who* has the responsibility for the doing. A procedure will normally call up any supporting instructions or standard forms.

The Instruction is a lower tier document which defines activities to be carried out, without defining discrete responsibilities.

Compilation of information

The data produced as described above should be carefully checked, authorized by the appropriate person in the issuing department and incorporated into a draft manual or manuals.

Introducing the system

Quality systems should not be imposed – they should be introduced on a progressive basis in a manner which generates the understanding, acceptance and commitment of all concerned.

When the draft information has been prepared, it should be tested and debugged. No matter how well we believe documents such as procedures have been written, they are unlikely to be

right first time. The way ahead is to introduce the draft information on a trial basis; three to six months is recommended, during
which time as many people as possible should be asked to work
to the system and identify and report any difficulties encountered
to their managers or other nominated person (maybe the person
responsible for coordinating the QA development exercise).

Substantial problems will need to be addressed and corrected
immediately. Others should be reviewed at the end of the test
period and the documentation amended as appropriate in light
of the user experience, subject to the agreement of the depart-
ment/discipline manager responsible.

Through this debugging stage we have got people involved,
provided opportunity for comment and shown willingness to
respond. We have hopefully encouraged commitment.

The system can now be introduced formally, by distributing
from a controlling source and within a proper distribution control
system, copies of the manual(s) – or, in some cases, procedures
only.

However, introduction should not be by imposition, but by
each manager/discipline head, meeting his/her staff and ensuring that there is a common understanding of management's procedural interpretation of requirements. (This step helps to fulfil
a responsibility placed on management under 4.1 of ISO 9001,
to ensure that the management system for quality attainment
is understood.) A record should be kept of such introductory
training.

System review

The introduction of the quality system is the beginning, not the end.

Once the system is installed it is essential that things are regularly monitored and reviewed to ensure that the system remains truly reflective of current technologies and working practices; also that all concerned are continuing to adhere to the system.

The two main review techniques used are:

- management review
- internal audit.

Management review requires that the company's executive management periodically and formally reviews the ongoing accuracy of the declared policy in relation to changing market and internal situations; also, that the quality system is proving adequate and effective in terms of meeting the management and control needs of the business.

Typical topics that would be discussed include:

- Are we continuing to identify fully with our customers' known needs?
- Is our policy still reflective of current market and internal requirements?
- Has the scope of our activities altered and, if so, has our quality system been amended to cater for the changes?
- Does our system continue to reflect the technologies/working practices in use?
- Have corrective measures introduced as a result of problem identifications, audits, etc proved effective?
- Have the forward training needs of staff been identified and planned for?

A formal record of such reviews must be established and actions monitored.

The internal audit

There are different types of audit which can be carried out internally to test the adequacy, understanding and application of the quality system. In a project-related industry such as the construction industry, the two dominant audit types are:

- internal system audits
- project audits.

It is not intended in this book to go into depth about the techniques for preparing, implementing and closing out audits. It is sufficient for present purposes to set out the features common to both types of audit and those characteristics which differentiate them from one another.

Purpose of audit

The basic purposes of an audit are to confirm that:

- there is an adequate management system in place
- it is available to and understood by those performing the tasks
- it is being properly adhered to in practice.

The perception of auditors by others and the attitudes of auditors to their task

Unfortunately, auditing is too frequently seen as a criticizing exercise. Many auditors seem to find satisfaction in recording trivia to justify the thoroughness of their audit. The good things found during an audit are often not acknowledged, while problems are exaggerated unnecessarily; as a result, the final report is often unbalanced.

An audit should be an aid to management, which both confirms and acknowledges proper implementation of the system's requirements, but points out any weaknesses found, enabling management to take corrective actions. Reports should be balanced and acknowledge good performances where they exist.

Who should carry out audits?

Audits need to be carried out by persons who are trained in audit techniques and who are independent of the area being audited. The scope of training needs to include such aspects as:

- preparing audit schedules
- arranging audits
- selecting the audit team
- gathering information
- team briefing and task allocation
- preparation and uses of checklists
- opening meetings and their purposes
- conducting audits, including questioning techniques, fact agreement and recording, behaviour and attitude, dealing with different personalities, thinking on one's feet, liaising with other team members, and so forth
- information evaluation and presentation
- closing meetings, their purpose and conduct
- preparation and issue of audit reports
- evaluation and agreement of corrective action responses
- corrective action follow up
- audit close out.

From the above it will be appreciated that there is much more to auditing than merely asking questions, looking at examples of work and taking notes.

Audit schedules

Internal system type audit schedules define the forward programme of intended audits and are usually calendar-based. Such a schedule might show that all disciplines within the organization would be audited once during a 12-month cycle. Project audit schedules similarly identify intended audits, but at key points in the evolution of the project, rather than on a strict calendar basis. Such schedules are normally agreed in advance between QA department and senior management concerning the systems audits, while agreement is with the project manager for the project audits.

Audit reports

These would go to the discipline managers in the case of the systems audits and to the project manager for the project audits and maybe to the senior executive.

Auditor performance

There are good auditors, and there are poor ones. Without dwelling on the poor, it is worth while mentioning a few of the hallmarks of the good auditor:

1. Checklists will not be visible; they need to be nothing more than memory joggers for the good auditor.
2. People under audit will not feel under pressure, because the auditor will have the ability to sense changes in auditees' attitudes, know how to relax tensions and still get the information sought.
3. Non-compliance situations will be fully discussed and agreed at the time they are noted.
4. Good points will be noted and acknowledged. Saying 'well done' to people when they deserve it reassures auditees that the auditor is fair minded and not purely engaged on a fault-finding mission.
5. At the end of the audit contact will be made with the head of department to give a resumé of audit findings. It may just be that the head of department can provide further information which may alter the flavour of a finding. (When the overall findings of audits are presented at the final meeting, there should be no surprises.)
6. When evaluating audit findings a good auditor will condense them if at all possible. For instance, if six examples of non-compliance are found of a similar nature, say regarding 'document control' in one work area, the good auditor will not proliferate paperwork by raising six non-compliance sheets. Instead, one sheet will be raised identifying document control as a general weakness in the area concerned, and the six examples will then be quoted.
7. Although auditors need to be independent (that is, not assessing their own departments), they do need to have a

sound understanding of the type of work going on, in order that they can ask sensible questions, appreciate what they are looking at and, when a non-compliance is found, be aware of its true significance.

Conveying and recognizing the audit message

The way audit findings are recorded is very important and it is the significance of what is found, as opposed to the detail, that usually needs to be conveyed. Similarly, when responding to audit findings the person taking corrective action should ensure that the real cause of the problem is being addressed and not merely an individual consequence.

For example, an audit finding may read: 'Code of Practice XYZ issue 3 being used in Design Office. This should have been issue 5.' A corrective action response such as: 'Issue 3 withdrawn and replaced by Issue 5' would not really be adequate. For example it doesn't question whether it was an isolated case or a symptom of a larger control problem. It doesn't question why or how the wrong document was issued. It doesn't question whether work that has been done contains errors, and so on. These points would have been followed through by a good auditor, and led to both a better defined non-compliance description and a more complete response.

Corrective action follow ups

It is important that follow ups are carried out promptly after the agreed date of the action implementation. If the action taken does not fully reflect what was agreed, then the consequence of not checking promptly may be continuing risk situations.

The selective use of the system

The philosophy of the QA approach is that it brings about cost-effective performance. This means that the quality system should not be applied like a sledgehammer to crack a nut, but selectively to the minimum degree necessary to meet the specified requirements of the job.

The normal way of accomplishing this is to assess the management and control requirements of the item or work package and assign to it a level of ISO 9001, 9002 or 9003, or other QA standard if applicable. The questions used for such assessments typically include:

- Could a part or structure failure lead to a safety risk to public or user?
- In event of failure is there good access for repair?
- In event of failure could there be serious process or cost implications?
- Does replacement or repair involve design input?
- Does replacement or repair involve process control measures?
- Can replacement or repair integrity be demonstrated by final inspection and test only?
- Is the item a standard, off-the-shelf unit, readily obtainable and of proven reliability?

The questions asked will vary from industry to industry; the key factors for a nuclear structure may differ from those of an off-shore structure.

Key factors should be agreed with the client, and appropriate weightings given when necessary. It is obvious that those factors which have safety and major cost implications need to be given appropriate QA considerations (eg to comply with CDM regulations). A structure having genuine design implications for those doing the work would be assigned a level of ISO 9001 whilst ISO 9002 would be stipulated if no design, but genuine process controls were involved.

It should be remembered, however, that items or work packages can be of commercial standard and not demand any formal quality system by the supplier or producer. The correct time for carrying out such evaluations is at the design stage. Those involved will normally be from design, QA and safety or purchasing disciplines.

Decisions should be recorded on suitable proformas, showing:

● questions
● answers
● those involved in the evaluation
● levels determined
● date of evaluation
● signature of authorization,

The above evidence should be held within the organization quality records.

This type of categorization, involving as it does safety implications, is very significant. Not only does it indicate the QA capability required of those to be invited to tender or to be awarded contracts, but the results of such categorization can also influence other important considerations, such as inspection frequencies, maintenance frequencies and calibration frequencies. They can also indicate the need for control documents such as quality plans.

Quality programmes and quality plans

There is a lot of confusion within industry between what are referred to as 'quality programmes' and 'quality plans'. Different terminology tends to be used in different industrial sectors. For example, in the nuclear industry, the term quality programme is used to describe what in effect is a quality system specific to a project or major contract. This would typically include:

- a description of the project scope of work
- project-related organization chart and supporting responsibility text
- reference to all applicable procedures, instructions and document formats
- a statement on interface involvements and limitations (eg client and consultant)
- a list of project-related work packages for which detailed quality plans are to be prepared
- a schedule of audits to be carried out
- a list of record documents to be produced.

The above requirements are not definitive and will vary according to project scopes. Such programmes are, however, tailored directly to cover the demands of the work in question and are therefore a unique statement of the management and control arrangements for the particular job.

In other sectors of industry (such as off-shore) such programmes may be referred to as quality plans or, more accurately, project quality plans (PQPs).

The main advantages of the quality programme or PQP are:

1. They are easy and quick to prepare, providing you have a formal quality system from which to select the applicable control procedures.
2. By having to study and include the scope of work, the chances of missing some important aspect are reduced.

Such a presentation can demonstrate at the tender stage both a good understanding of job requirements and an ability to manage the work programme in a competent manner. It can be an important influence and provide the tenderer with a true competitive edge over those who make lesser responses. It should be remembered that the tender stage is the most important. Failure at this stage means no contract.

The limitation of such programmes or PQPs is that, although they define who is to do what, which procedures are to be applied and the records needed, they fail to provide the degree of detail control upon which ultimate success really depends. The way to meet this need is by use of another document, which has been widely used in the manufacturing industry for many years and has recently been used more and more for the control of site activities in civil and structural work programmes. This document is the Quality Plan. To avoid confusion, this will henceforth be referred to as a *Detail Quality Plan* or *DQP*.

The DQP is used to control activities on discrete packages of work, of which there may be many within the total scope of a project. Typical packages could relate to:

- designing a structure
- fabricating a structure
- sinking piling
- making reinforced concrete sections
- erection of formwork
- placement of cladding
- manufacture or installation of a H and V system

DQPs fulfil three essential functions:

1. They demand thought about the task to be undertaken. This is because it is necessary to read relevant specifications, standards, instructions and so on, and to restate significant requirements on a different document – the DQP.
2. They enable visible controls to be exercised at every stage of the work programme.
3. At the end of a job the signed off DQP, along with the supporting documentation generated during its implementation, provides a total record of work integrity.

TITLE:-   QUALITY PLAN FOR: ....

PROJECT/JOB REF. ....

PREPARED BY: ....   DATE: ....

APPROVED BY: ....   DATE: ....

CLIENT APPROVED BY: ....   DATE: ....

PLAN BASE DOCUMENTS

DRAWINGS	ISS	CODES OF PRACTICE	ISS	STANDARDS	ISS	OTHER

PLAN COMPLETION CHECK LIST	SIGN OFF DATE
ALL DOCUMENTS ISSUED TO PLAN	
ALL VERIFICATIONS CARRIED OUT TO PLAN	
ALL TESTS COMPLETED TO PLAN	
ALL RECORDS PRODUCED TO PLAN	
ALL ACTIVITIES SIGNED OFF	
CLIENT ACCEPTANCE	

PLAN CODES   H = Hold Point   C = Document for client records
             A = 100% Inspection   D = Document for own records
             B = Final Inspection

Figure 4.5   Typical Detail Quality Plan front sheet

1	2	3	4			5	6		
		DEMANDING/	VERIFICATION				ISSUED TO		
OP. No.	ACTIVITY	CONTROLLING DOCUMENT(S)	CODE	BY	DATE	VERIFYING DOCUMENTS	CODE	SIG.	DATE
1	SUBMIT QUALITY PLAN FOR CLIENT APPROVAL	$SPEC^N$ A311	H			APPROVED QUALITY PLAN	CD D		
2	AGREE TECH. QUERIES	TQ 326	H			FORMAL CLIENT RESPONSE	CD		

CONTINUATION SHEET No. 2 OF 4

QUALITY PLAN No.
PREPARED BY
DATE

Figure 4.6 Typical Detail Quality Plan continuation sheet

Figures 4.5 and 4.6 show typical formats for the front sheet and continuation sheets respectively. Formats will vary from company to company. It should be noted that the front sheet contains provision for acceptance/approval by the client or contracting authority for whom the work is being undertaken before work actually begins. This not only allows clients to see that specified requirements have been fully identified and planned for, but also allows for them to insert codes (see bottom of front sheet) indicating those stages where they wish to have direct involvement, receive records and so forth.

The typical front sheet includes a list of key reference documents applicable to the scope of work covered by the plan. It will also usually contain provision for signifying that all key requirements of the plan have been totally met; for example 'All records produced to plan'.

The continuation sheet(s) of the plan are the actual controlling documents. They are usually divided into columns, each presenting key identification, control or record-related data. They show the way the plans are structured by identifying all important steps (activities) to be undertaken to satisfy the key requirements of the contracts, specifications, codes of practice, procedures, drawings and so on.

Activities are presented in sequential order.

Column 1   shows the activity number
Column 2   describes the activity
Column 3   identifies the document(s) which require the activity to be done and which specify the means
Column 4   identifies the level of verification or control to be applied during the activity. This will be signified by one or more of the code numbers listed on Sheet 1 and can include codes representing involvement by either or both parties to the contract who will append their signatures of stage acceptance when the activity in question is satisfactorily accomplished.
Column 5   states what is to constitute evidence of the activity in column 2 being properly completed in accordance with the controls defined in column 3

Column 6   defines where the records identified in column 5 are required to be distributed – client, contractor, and so on. This information is usually expressed in the form of code numbers as shown on Sheet 1.

It should be appreciated that it is often possible to produce generic DQPs which can be used for different contracts, because activities such as the preparation of precast reinforced sections tend to be carried out in a standard sequence. It is then only necessary to alter details such as client name, product title and number and client involvements for each contract.

5 Quality Assurance and the Client

Introduction

Within the construction industry there is the well-established chain of client, consultant and contractor. Sometimes the construction works are supervised on behalf of the client by the consultant, sometimes by the client direct, depending upon skills and resources available. For a project to be successful, however, all must play their part and therefore understand clearly what their part is. To enable this to happen it is essential that there are well-defined interfaces, clear levels of responsibility and properly specified requirements, including those concerning work planning, resourcing, skills, acceptance criteria, records and so on.

All too frequently, clients fail to provide proper information and guidance to those from whom they expect design and construction support. Positive direction should be expected from and provided by a client. The results of poor direction are vague briefs, inadequate specifications, disproportionate numbers of post-contract amendments, changes, concessions, variation costs and final results which often represent compromise rather than

71

the achievement of real requirements. As they are commissioning the work, clients tend to consider themselves to be the prime customer. But, while this may be the case, clients must also recognize their obligation to consultants, contractors and suppliers who are, in effect, 'customers' requiring the proper information and guidance needed to carry out the work satisfactorily. In other words, there needs to be a better general awareness that our customers are those who need and receive outputs. In the case of the client, that can mean everybody, directly or indirectly, at some time and in some form.

A typical example of the need for clients to prepare better information is recognized to a degree under the clients' duties as described in the 1994 CDM Regulations concerning health and safety in construction.

The regulations require the client to provide the appointed Planning Supervisor with information concerning such matters as:

- site information
- existing structures
- access limitations
- water courses
- utilities and so on.

In order to assist the latter prepare stage 1 of the pre-tender Health and Safety Plan.

The construction industry only really began to come to grips with QA in the early 1980s, although QA techniques have been successfully applied in many other sectors of industry for over 30 years and the benefits are widely recognized.

The significance of quality management for the client cannot be ignored. Even if a client does not have QA or impose it upon others, its effects will not be escaped.

Many consultants and contractors, faced with the need to tender competitively, have now set up Quality Systems, realizing that sensibly applied QA will enable them to perform more effectively. One requirement of the Quality System will be that they carry out contract reviews which will increasingly force clients who seek to engage them to provide definitive specifications from the outset.

The QA aware client

Take a hypothetical situation. The client is a large organization with its own Building, Civil and Structural disciplines, plus supporting departments such as Procurement, Inspection and Quality Assurance. It has a well-documented quality system, meeting the relevant requirements of ISO 9001. The organization intends to undertake an important design and build project, involving civil and structural design, some off-site steelwork fabrication and a big on-site build programme. The client owns the site and will be the end user of the finished building.

It is proposed that an in-house Project Team, comprised of selected personnel from all key disciplines, should manage the job. The bulk of the design work will be done by consultants. The client will select all off-site and on-site contractors and suppliers, and manage site work through its own Resident Engineers Department.

Let us look at when, why and how QA would apply on such a project and how important having a quality system can prove to be.

Support to the Project Team
This includes a nominated Quality Engineer, with specific duties relating to the application of QA on the project. These duties are agreed with the Project Manager and include:

- help with preparation of a project manual and selection or adaptation of procedures
- control, issue, update and so on of the manual and procedures, as requested by the Project Manager
- guidance on Project Quality Programmes and/or Detail Quality Plans
- setting up and ensuring implementation of project specific audit schedules (internal and external)

- evaluations/assessments of QA capabilities of potential consultants and contractors, and liaising with same to guide them in setting up their quality programmes
- ongoing support and audit of chosen consultants and contractors
- setting quality levels for various work packages in order to ensure correct quality conditions are included in tenders and contract documents
- QA inputs to tender and contract documents
- advice on adequacy of QA responses received in tender submissions
- periodic reviews of quality documentation such as procedures
- provision of training on QA matters to Project Team
- representing Project Team in liaison with others on quality matters.

If we consider the above duties within the framework of the project timescale, we see they form a logical pattern of management and control throughout the scope of work.

The first task is obviously to identify how the project is to be managed, what the individual responsibilities of people on the Project Team are going to be and how functions and people relate. This is a fundamental requirement of ISO 9001 and the simplest way to fulfil this need, is by means of an organization chart and supporting text (see p. 90).

The second need is to identify the procedures by which those working on the Project Team will carry out their specific tasks. It is a question of selecting from your total menu of procedures those which are needed for this particular job. (It is obvious that this would not be possible unless there was already a proper system in place.)

The third task is that of compiling the O&R information, procedures and other basic data into a Project Manual, for authorization by the Project Manager.

One procedure in the Manual will concern the controlled issue, recovery and change of the Manual and procedures themselves. This function is accepted by the QA Engineer on behalf of the Project Manager in this instance.

Following preparation of the Project Manual, its implications are discussed with all members of the Project Team, ensuring that everyone has a common understanding and interpretation of it. As is usual this exercise is carried out by the Project Manager or appropriate nominee and QA Engineer. Issue then takes place in a controlled manner, in accordance with the relevant procedure.

As the job develops and the need to select consultants, contractors and suppliers arises, the various work packages are assessed and levels of quality determined (see pp. 63–4) Consultants and contractors are then invited to bid for the work, submitting with their bids evidence of their QA capability (normally in the form of an Outline Quality Programme or plan).

Submissions are assessed from the QA standpoint by the Quality Engineer, who seeks further evidence, if necessary, of QA capability and then advises the Project Manager of the results.

During implementation of the work programmes, the client's QA Engineer carries out a number of functions, as follows:

1. Works with those consultants and contractors for whom the client has assumed direct responsibility, to ensure that their Quality Programmes and/or Quality Plans are comprehensive and being properly applied.
2. Ensures that audits are being carried out on the activities of the consultants and contractors by their own QA representatives.
3. Liaises continually with consultants and contractors to ensure interface problems are quickly and properly resolved.
4. Carries out both internal and external audits to confirm ongoing adherence to agreed arrangements and to ensure that any corrective actions are promptly taken.
5. Reviews and maintains the project manual and procedures.
6. Finishes with a final review of hand-over records.

It will be seen that all the activities described in this example are covered by the quality system.

The QA unaware client

A client organization not practising QA should perhaps ask itself a few basic questions, namely:

1. How much does blind delegation cost?
2. How well managed and controlled are selected consultants/ contractors?
3. How effectively and efficiently do they interface?
4. How much are their shortcomings costing?
5. Why are there adverse closed order variances on projects?
6. How much time is spent in hassle, recriminations, litigations and how much of this is due to own shortcomings?
7. When things go wrong, what lessons have been learned (other than to go elsewhere for support next time)?
8. How much is spent on own management and control (QA) awareness and capability?

In many cases clients can and should do far more to protect their own interests, that is to get a finished result of the right quality on programme and to cost. We would recommend they consider investing in some QA training of key staff to ensure that:

- internal activities are effective, efficient and being properly managed
- consultants and contractors are realistically chosen and controlled
- interfaces are properly managed
- visible, planned, ongoing controls are exercised
- better fault analysis and cause eradication is carried out
- proper records are produced

Or such organizations should engage suitable support in the form of additional QA staff or an external QA consultant.

Quality Assurance is not restrictive. It is a powerful management tool. Ignoring or discounting its advantages will not alter the fact that it is being increasingly accepted worldwide. By not adopting it you may be limiting your own potential, profitability and ability to compete.

6 Quality Assurance and the Design Team

Introduction

The need for quality assurance within the design team should now be evident. The method by which a streamlined and appropriate system is established for the design team might, however, be less obvious. The procedures most appropriate to any company are those that are developed by and tailored to meet the specific requirements and nature of the company and will to a great extent reflect working practices already in operation. This chapter, therefore, sets out the basic principles associated with the development of working procedures for design and gives guidance on typical procedures that might be developed in specific circumstances.

It is important to note that the following recommendations should *not* be read in isolation from other parts of this book.

Chapter 6 is complementary to Chapter 4, and any overlap between them is intended to emphasize the important aspects.

The system development team

What then do you do at the outset, when faced with the daunting prospect of preparing a quality system? We would suggest that before you put pen to paper and develop the system, you undertake the following:

1. Establish a team within your organization who will be responsible for preparing the QA documentation. This team should preferably be led by the individual who will ultimately be responsible for managing the system (the QA manager).
2. Establish within that team a clear and common understanding of your current working practices. In addition, establish which of those current working practices are efficient and effective and which are not.
3. Decide which quality standard you will be complying with (most likely ISO 9001).
4. Establish the most efficient format for preparing the procedures and ensure that all members of your system development team are given the appropriate training, both in the principles of quality assurance and in the process of procedure writing.

It is often useful, at this stage, to engage an external adviser to give initial training to the team and guidance on achieving its objectives most effectively. However, the responsibility for deciding which procedures will be developed, and in what form, rests with you.

Who should be invited to join the system development team? The simple answer is those individuals who will be immediately responsible for implementing the procedures when you launch the system. Clearly, it is they who will be the ultimate arbiters of success or failure of the system and they should be drawn into the system development as early as possible, in order to ensure their personal commitment. The team should always be a manageable size, and in larger companies and public organizations it may be appropriate to approach the procedure writing on a departmental basis (see pp. 56 and 83).

Establishing the system structure

Current working procedures can be set down by requesting the members of the system development team to state, clearly and succinctly, what is happening at design team level. It is important at this stage to refer to the quality standard to establish what areas are to be covered (see Chapter 3).

For instance, the team leader may request the chief engineer to state the current design control procedures or practices within the organization, or a principal architect may be asked to state the manner in which incoming, design-related documentation is recorded and processed to the appropriate desk or file.

However, we would suggest that before this exercise is started, the team should make a clear statement of the typical manner in which the business is conducted – the chronological stages by which a typical project is received and dealt with in the office. A typical example for a firm of consulting engineers is given here:

1. Client invites Senior Partner to a meeting.
2. Senior Partner instructs Associate to make a preliminary survey of proposed site.
3. Senior Partner and Associate discuss potential schemes.
4. Associate Graduate and Technicians prepare concept designs.
5. Senior Partner and Associate discuss with Client the three schemes. Client selects two for detailed evaluation.
6. Associate's team develops both schemes in outline general arrangement and supervises preparation of approximate quantities and prices.
7. Associate prepares report for client, setting out the broad capital commitment for each scheme alongside the merits and demerits.
8. Client commissions detailed design of one scheme.
9. Senior Partner augments team by appointment of complementary professional advisers.
10. The team, led by the Associate, undertakes the following actions:

a)   commissions subsoil survey
b)   updates general arrangement drawings
c)   prepares engineering calculations for strength, stability and serviceability
d)   prepares detailed construction drawings
e)   prepares specification for construction works
f)   prepares bills of quantities for the construction works
g)   selects contract conditions for the construction works
h)   prequalifies construction contractors
i)   issues tender package
j)   negotiates statutory certification
k)   presents client with tender assessment
l)   prepares contract forms for signature by client and contractor
m)   supervises contract on client's behalf and certifies payment
n)   prepares completion certificate
o)   prepares 'as built' drawings
p)   archives the project.

In most design companies commissions are, of course, received in different ways. It is important, therefore, to incorporate in this list all the different steps that can be envisaged. It may be sensible to set down two or three chronological sequences in order to accommodate all the possible variations in a sensible manner.

Let us assume, for the purposes of this exercise, that the above chart incorporates all the steps typically undertaken by your company. How then do we now design the structure of the system to accommodate all these steps?

Firstly, the chart provides a list of the principal individuals associated with the development of the design. It is necessary to provide definitions of responsibility for these individuals, and to determine their position within the hierarchy of the company.

From sections 1 and 8 it is clear that a contract exists between your design company and the client. Formal quality assurance demands that the terms of that contract are fully understood by both parties, and a procedure is indicated, therefore, for contract review. This will ensure that the client's requirements are always fully understood.

Sections 2 and 10 suggest that your company requires a procedure for site surveys. Sections 2 to 7 suggest the necessity for a procedure on concept design. This procedure will also cover, in broad terms, the requirements for the client report.

Section 10 incorporates 16 individual fields of activity which throughout the period of the project will be overlapping and interactive. Having committed resources to the development of the project, the client will not wish to suffer any delay in commissioning, so the activities in section 10 must clearly be compressed into as short a timescale as possible.

The need for a *modus operandi* is therefore obvious. Formal quality assurance requires that the activities in section 10 are controlled by a formal Project Quality Plan. The authors believe that the development of efficient Project Quality Plans are essential to successful quality assurance for the design team. This subject is covered in greater detail later in this chapter.

The customer may wish to be involved in the development of the design. It is also common in many projects to appoint external professional advisers, as in section 9, to augment the project team. A procedure may therefore, be required to control the flow of data from external parties, to ensure that the design is always developed on current information.

Section 10 clearly indicates a need for a procedure or procedures for design calculations and drawings which must ensure that the design base information utilized by the team is current and appropriate, that the use of computers is safe and verifiable, that the issue of documents is thoroughly controlled and that the process of change can be accommodated efficiently and thoroughly. Clearly, this is a key area to be addressed by the system.

Section 10 (f) requires that a procedure is written for the preparation of bills of quantities or, alternatively, the control of sub-contract services for the provision of such effort.

Section 10 (h–n) shows that a procedure is needed for the control of contracts on site. This requires a complementary procedure on the method of placing contracts.

Finally, section 10 (o and p) necessitates the development of a procedure to ensure that feedback from the construction process is adequate and that the archives reflect a fully verified set of contract documents.

It is highly unlikely that the above analysis will yield a complete set of procedures for the company operation. However, it is possible that, taken in association with an analysis of the requirements of the QA standard, it will provide the necessary list of procedures to comply with the requirements of formal quality assurance.

The sort of procedures that might be added as a result of an analysis of the standard are 'control of the technical library' and 'preparation of letters, reports and so on'. In addition, your quality system will require a procedure on the review of the QA system and the preparation and issue of work instructions.

Having completed this exercise, the process of determining current working practice will be far more structured and hence economical in effort.

You are now in a position to write down a list of procedures and determine the structure of the system. This will probably change in the process of development of the system, to reflect your increasing understanding of the requirements of formal QA; however, it is an essential starting point. We would suggest that, as far as possible, the structure of the procedural section of the manual reflects the chronological sequence you have already established.

As stated at the beginning of this chapter, we are not attempting to provide a complete list of procedures for any one practice. However, by providing a typical menu, we may cover a substantial proportion of your requirements and enable the demonstration of a typical approach to developing the system. The following list is indicative only, therefore, of the nature of the list that you will compile. Likewise, you may decide that two or three from the list may be combined into a single procedure or that a more comprehensive split of titles is appropriate. The following list assumes that you are involved with architecture, engineering design, quantity surveying and site management. (*It is complementary to Chapter 4, which must be read to give the full breadth of system requirements.*)

Sample system structure

Part 1
1.1   Policy statement
1.2   Company organization and responsibilities
1.3   Training (this may be included in the Procedures)

Part 2 – PROCEDURES
2.1   Contract review and concept designs
2.2   Preparation of a Project Quality Plan
2.3   Control of data received
2.4   Control of design base information
2.5   Design control
2.6   Preparation of detail drawings
2.7   Contract cost measurement
2.8   Control of records
2.9   Preparation of specifications
2.10   Control of the library
2.11   Concessions and corrective actions
2.12   Review of the QA system
2.13   Placing contracts
2.14   Writing and controlling work instructions
2.15   Control of subcontract services
2.16   Equipment calibration
2.17   Document issues
2.18   Use of computers
2.19   Contract management
2.20   Contract completion

It is now possible to split the list between your system development team, requesting that they write down a brief description of the current working practices associated with each activity. Your working party may then decide whether improvements are possible prior to the preparation of the procedures.

Remember that your staff will not have the time to wade through masses of paper, so keep the sections as short as possible, splitting the content into as many different procedures as is feasible.

We can now address each of our sample sections and procedures. Since the policy statement has been covered at (pp. 45–6),
we will start with 'Company organization and responsibilities'.

Company organization and responsibilities (1.2)

It is generally accepted that good human relations and communication, with clear definitions of responsibility, are essential in the
development of any management system. The importance of
defining the company management structure cannot be overemphasized, and clear limits of responsibility and authority
must be given for each level of staff involved in the process of
design and project control.

When developing the 'Company organization and responsibilities' section, avoid defining duties, as doing so prevents
delegation and flexibility in the management structure. All that is
required, within this section, is a statement of the basic responsibilities allocated to each level of staff and the line of reporting.
An architectural design practice might for instance specify the
responsibilities of its associates and senior architects thus:

> *Associates*
> The Associates are responsible to the Partners for the pro-
> duction of design and contract documents in accordance
> with the Quality Assurance procedures and customers'
> specified requirements. The Associates may undertake
> responsibilities for a number of projects and, at the discre-
> tion of the Partners, individual project finance, administra-
> tion and office management.
> The key function of the Associates is to act as Senior
> Architects on the major commissions undertaken by the
> practice. Reference should, therefore, be made to the
> responsibilities of the Senior Architects.
> *Senior Architects*
> The Senior Architects are responsible to the Associates and
> Partners for the production of architectural schemes, design
> and detail in accordance with instructions from the

Associate and the Quality Assurance procedures. Where deemed appropriate, the Senior Architect may be required to undertake responsibility for individual project management and advise and assist the Associates and Partners on project finance and administration, specifically related to their individual workscope.

It may be necessary within this section to state the broad objectives of the practice with regard to minimum qualifications for each level of staff. Again, it is essential that such a definition does not prevent the company executive from making appointments justified by ability.

Quality assurance manager

Included within the company structure will be an individual who will be a member of the company's management and responsible for the management, on a day-to-day basis, of the quality system. That individual should, preferably, not be involved in the immediate project management and should report directly to the chief executive or senior partner. This is not always economically feasible or indeed practical, and great care should, therefore, be given in such circumstances, to choosing an individual who is capable of taking a dispassionate view in difficult circumstances. It may be appropriate in a smaller organization to appoint the Quality Manager as a part-time position, albeit in the initial stages of setting up the system it could involve a considerable amount of the individual's time.

The organization and responsibility section should be accompanied by a chart indicating the lines of communication and responsibility within the company. This 'family tree' will become a very sensitive document and should be as succinct as possible, avoiding the temptation to nominate individuals against the different levels specified.

Training (1.3)

A key element of formal quality assurance is visible commitment on the part of the organization to ensuring that all members of staff performing quality related activities, are suitably qualified and trained. It is therefore incumbent upon the organization to devise and implement a comprehensive training scheme which reflects the requirements of the marketplace in which they operate. In addition, as with many aspects of quality assurance, it is necessary to keep records for each member of staff. This section must cover training in the QA system and may be included as a procedure.

Contract review and concept designs (2.1)

The importance of establishing clear definitions of scope and responsibility has already been established. At concept design stage of the project, the opportunity for confusion is substantial, and we would suggest that a procedure is developed to establish clearly the requirements of the client.

It is not uncommon for the client to issue verbal instruction, in the first instance, at a contract initiation meeting. The procedure should, therefore, establish a method by which the client's requirements are defined, recorded and confirmed. Your procedure may indeed list specific aspects of the agreement, without which the definition of your work-scope is incomplete. That list might include such items as:

- project scope and outline programme
- client quality assurance requirements
- codes of practice and standards to be adopted
- statutory bodies to be satisfied.

Your procedure will almost certainly define the process by which such fundamental aspects of the project are defined and understood. This may be in the form of a meeting between the principal staff member who received the client instruction and the senior staff member who will be ultimately charged with the responsibility for the project. The minutes of this meeting would form the basis of the confirmatory letter to the client. The formal confirmation will also request notification of the client's principal contact for the project and confirmation of the consultant's principal contact.

The establishment of a procedure for conceptual design is extremely difficult, since the designers involved must not be constrained in their creativity. However, you may deem it appropriate to establish a format for regular review of the designs, preferably including the client's representative, to ensure that the principal objectives of the project are being met. Clearly, also, a system should be established for presenting concept designs to the client in a manner whereby the client is able to make the correct choice on which to proceed with the fully detailed design. It may be appropriate, therefore, to incorporate within this section the skeleton structure of a 'conceptual design report' appropriate to your practice. A model for such a structure is provided in the publication *Civil Engineering Procedure*, published by the Institution of Civil Engineers.

As with all design processes, the control of input and output data must be specified, and the provision for comprehensive records established.

Preparation of a Project Quality Plan (PQP) (2.2)

Following agreement of the conceptual design to be adopted for the project, the business of preparing the contract documents can start in earnest. A contract will now exist between the consultant and the client for the development of the project and any good management text will demand that the contract is clearly understood by all parties associated with it.

The PQP is designed to establish a firm understanding between the client and the designer of the relationships and scope of work embodied in the contract between them. It should clearly define the processes of management to be utilized by the design company to ensure that the project will be progressed in an efficient and thorough manner.

There are various guideline documents available which specify minimum requirements for a PQP. The following is a list of typical requirements, but it may not be comprehensive for your company:

- scope of work
- project organization and responsibilities
- project management
- client project review, verification and approval
- project quality procedures
- design basis and programme
- project phasing
- design reviews
- records and retention period.

This document should be compiled at the earliest opportunity and it may be appropriate, in the first instance, to issue an abbreviated plan whilst awaiting confirmation of undefined sections.

As previously stated, we consider the establishment of a good PQP is the keystone on which successful quality assurance is established. It is therefore worth reviewing in a little more detail each of the above mentioned sections.

Scope of work

The client will not always supply a formal written scope of work to the consultant prior to the commencement of project document preparation. In addition, where the client has provided the concept design, the initial instructions from the client may be to commence the development of detail design immediately.

The extent of the contract which exists, therefore, between the client and the designer should be firmly established, and the designer should state his understanding of the activities required to complete his scope of work and formally record this in the PQP.

As with all sections within the PQP, we must recognize the possibility that the management of the project might change within the contract period. The establishment of the basis of contract in a formal manner is essential, therefore, to permit another senior manager to take control if necessary.

Likewise this section, as with others in the Project Quality Plan, provides an essential basis of understanding and communication for the other members of the design team. It should, therefore, be incumbent upon the consultants' representative to ensure that all members of the design team are given the opportunity to study the PQP.

Project organization and responsibilities

The broad definition of company organization and responsibilities will already, have been included in part 1 of your system. However, in order to define the specific responsibilities of the project personnel, the PQP should identify those individuals responsible for the project line management.

This is obviously project specific and accomplishes two objectives. Firstly, for good communication and continuity, the individual members of the design team need to be appraised of the overall management structure; and secondly, the client requires to know the identity of key personnel in order to maintain efficient communication between the parties in contract. This section may simply comprise a line management chart like the example in Figure 6.1. Clearly, where the services of other professional bodies are employed to augment the design team, additional descriptions of responsibilities may be necessary.

Project management

On some engineering projects, the role of project manager is provided by a client representative. In this event, it will be necessary to provide, with the PQP, a definition of the limits of responsibility for the various individuals, including the client representative. Again this may be accommodated within a management chart, or key relationship chart, as in Figure 6.2.

Client project review, verification and approval

In order to maintain efficient lines of communication, formal quality assurance requires that the PQP describes the arrangements for the client to review, verify and approve the design at the appropriate milestones in the programme. Such arrangements should be defined within this section, along with a schedule of agreed dates for review.

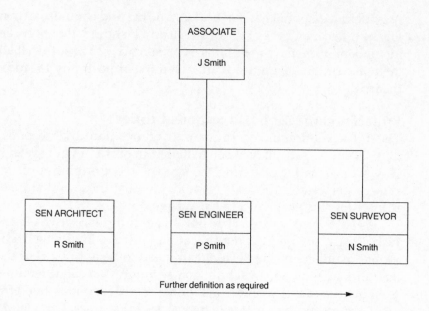

Figure 6.1   A line management chart

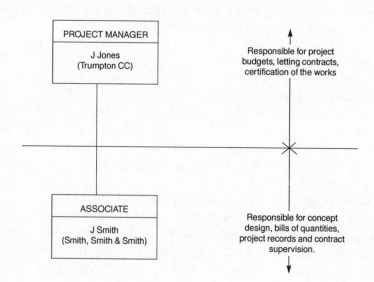

Figure 6.2   A key relationship chart

As with many aspects of quality assurance, this section merely serves to formalize the process of communication and approval such that misunderstanding and abortive effort may be avoided.

Project quality procedures

This section of the PQP lists the standard procedures applicable to the project as referenced in the company's system manual. In order to avoid an unnecessary proliferation of paper, it is important to recognize that any particular project may not require the application of all the procedures referenced in the company's quality system. This section defines procedures which are applicable to the project under consideration and is, therefore, a useful method of preventing unnecessary paperwork; that is, the application of QA by exception.

It should be noted that this is only possible if the system manual permits the individual quality related activities to be independently described in separate procedures, and it is further evidence, therefore, of the necessity of a well-structured list of procedures.

Design basis and programme

Design basis refers to all information, specifications and standards upon which the design will be based. This procedure must demand an examination of the correct revision to documents and standards to be utilized on the project, such that a design can proceed in a coherent manner. The procedure may also demand specification of standard references such as books and papers, depending on the policy of the company.

The design basis should also include reference to all client specific documentation and technical reports which will be required before the design can be progressed. Typical examples of such information are process or operating conditions, loading information and site soil data.

In addition, it is also a requirement of formal quality assurance that the PQP incorporates a programme for the design. The input dates of all key design basis documents should be specified as milestones on the programme, as well as construction start date and tender completion.

Project phasing

The planning of resources within a multidiscipline project is one of the more difficult aspects of any project management function. Invariably, the project programme is governed by the release of information from external parties, without which the design cannot be progressed.

It is, therefore, extremely useful at the commencement of the project to set out the requirements for staff resources in a bar line programme or histogram which then becomes a live document and can be adjusted to suit as the project proceeds.

Design reviews

One important aspect of formal quality assurance is that design reviews are held, recorded and acted upon.

The design review would normally be a structured meeting between the company senior management and the design team representatives to discuss the principles of the design in a thorough manner and to ensure that there is no deviation from the client's requirements, either at the outset or during the continuing design process.

Design reviews offer a useful vehicle for the company senior management to be involved in the immediate process of design and also for the design representative to take a fresh look at the principle decisions governing the process of project development. In addition, with regard to the training of junior members of the team, such meetings offer an excellent opportunity for involvement in the decision process and to witness the process of problem resolution. It is feasible to combine this requirement with the requirement for client project review in order to save time.

Such reviews should be scheduled in the PQP along with a schedule of quality audits, which will be incorporated into the overall company audit schedule. Audits will be undertaken to establish that the project procedures are being applied in a thorough and constructive manner and to identify problems with the quality system. Clearly, audits should be carried out by a member of the company's staff not directly involved with the project.

Records and retention period

Another essential aspect of formal quality assurance and a PQP is a statement of what records should be kept during the process of the project and the periods for which records should be retained. A structured project filing system is helpful in this respect, and it may be that your system would benefit from a separate procedure on the maintenance and control of records (see p. 102).

Requirements for archiving records are also often determined in a separate 'Contract completion' procedure (see p. 107).

It is, however, common for the client to wish to retain the principal project records, and this section of the PQP should identify which records will be handed over upon project completion and the period for which the design organization will retain copies. With regard to design liability, it is important that this period of archive retention is reviewed by senior management on a regular basis.

If the design organization (for example, the consultancy) is also providing the Planning Supervisor service (re CDM 1994 Regulations), then one important record that will need to be compiled for ultimate handover to the client will be the Health and Safety File.

Clearly, much of the above information may be subject to change throughout the period of the contract. The establishment of the PQP enables the project management to review the necessity for change. Amendments to the PQP will be issued when deemed appropriate, as a result of this review.

Control of data received (2.3)

The control of data received is an essential element of formal communication and efficient project management. Data may be received by various means:

- letters, telex, memoranda, telefax or e-mail
- reports from technical advisers
- design calculations from technical advisers
- comments sheets (based on company documents)

- working drawings from the client, technical advisers or sub-contractors, noting design approvals, changes or working details
- telephone conversations
- meetings and discussions.

The quality system must describe the methods by which all incoming data are recorded, processed and filed in an appropriate manner.These procedures will obviously vary from company to company but they must be sufficiently structured to ensure that any document or instruction received can be accessed by the design team in an efficient manner.

It is often the case that instructions are received in an informal manner such as by telephone or 'round the board' discussion. In this event, the system should specify the method by which such instructions are recorded and transmitted back to the originator.

A common source of error within the design function is the misinterpretation of verbal instructions. It may, therefore, be appropriate to consider the use of a 'confirmation of a verbal instruction' (CVI) form in order to avoid confusion and bad feeling between client and design team.

The use of standard forms can greatly benefit the process of formal communication, and if the CVI is transmitted in duplicate such that the originator is obliged to acknowledge receipt, the verbal instruction is converted into a written one. The use of such forms has been found greatly beneficial by those who have already adopted formal quality assurance. A typical example of a CVI for the design office is given in Figure 6.3. It can be seen from this example that a statement of the implications of the instruction can greatly benefit a client's representative who may not have been aware of the wider effects of his instructions. In addition to the above, as is the practice with most design companies, it will also be necessary to establish a system for recording the discussion at meetings. There is nothing unusual in this requirement.

On large or multidisciplinary projects, it may be useful to establish, at an early stage in the project, a list of approved signatures, so that the incoming data or instructions are seen to be authorized in an appropriate manner.

CONFIRMATION OF VERBAL INSTRUCTION	DATE RECEIVED: TIME RECEIVED:	
ORIGINATED BY:	PROJECT No:	SHEET No:
RECEIVED BY:		
PROJECT TITLE:	TELEPHONE	
	FORMAL MEETING	
	INFORMAL DISCUSSION	
SUBJECT:		

INSTRUCTION:

IMPLICATION:

DISTRIBUTION: ORIGINATOR OF INSTRUCTION
ORIGINATOR OF FORM
DIRECTOR
OTHER

Date _____ Signed _____
Name _____

Receipt acknowledged _____ Date returned _____

YELLOW Customer Copy PINK Acknowledgement Copy WHITE File Copy

Figure 6.3   Example of confirmation of verbal instruction (CVI) form

Control of design base information (2.4)

Data control is an important aspect of any formal quality assurance system. Design base data is generally provided in the following documents at least:

- the British standards and codes of practice
- building regulations and other national/local statutory regulations
- international standards and codes of practice
- institution publications such as 'proceedings'
- learned body publications such as BCA or academic institutions
- customer standards and guidelines
- internal standards and guidelines.

A formal system of recording such information, and maintaining it such that it is comprehensive in terms of revisions and amendments, is essential. There are numerous systems on the market which automatically update British standards and European standards. It may be cost effective to install such a system; they are generally provided on microfilm or microfiche. Using outdated standards can be very detrimental to good quality assurance.

It is also necessary for the quality system to describe a method by which amendments, corrections and deletions to standards and other design base data are distributed to the project teams.

Design control (2.5)

This is obviously one of the key procedures within any system for the design office. Whilst we cover this subject in one section here, it may be appropriate in your organization to split it into a number of different headings, such as:

- scheme preparation
- design calculations, verification and approvals

- document issues
- change control
- feedback (concessions and corrective actions).

It seems reasonable, therefore, to review the requirements of design control under these headings.

Scheme preparation

This section will cover the preparation of architectural scheme design and general arrangement drawings (GAs). It will describe methods by which the scheme design and GAs are verified against the conceptual design and the design basis documents. Since the schemes will become the main reference for the preparation of the detail design drawings, it may be deemed appropriate to identify specific methods by which the philosophy and dimensional integrity are verified. It may also be deemed appropriate to specify within this section an approval stage prior to the commencement of detailed design.

Design calculations, verification and approvals

The necessity for the preparation of a procedure for design calculations is obvious. It is perhaps the one area where formalized procedures are already in operation in many consulting practices. This is evident by the wide usage of standard forms for the preparation of calculations which will normally require the following as a minimum:

- unique project reference number
- project title
- reference to elements under examination
- date prepared
- name of originator
- signature of verifier
- signature of approver
- page number
- issue status.

However, such a system is not entirely comprehensive, since it will not give sufficient information to describe the history of the calculation set. It may be appropriate to consider the introduction of a title sheet for calculations which would incorporate the following information:

- name of senior representative responsible
- name of originator and date of completion
- client's full name
- project title
- internal project number
- calculation content
- status, ie schematic or conceptual, preliminary, tender, working or other
- number of sheets in the calculation set
- name of verifier and date that the verification was completed
- signature of representative approving and date approved
- issue and revision record
- comments such as reasons for revisions.

An example of how such a sheet might be set out is given in Figure 6.4.

Formal quality assurance requires that verification of calculations is carried out in a thorough and visible manner. Your procedures should, therefore, specify that the project files contain a verification copy of all calculations, marked up to indicate that all numerical calculations and design assumptions have been independently verified.

It is useful to review methods of verification and to recognize that, for complicated calculations and analysis, alternative methods to those employed by the originator are useful to ensure that the calculations are correct. In addition, where testing is carried out to verify principles or material properties used in a calculation, a test report should be appended to the calculation set. In the event that the verifier and originator disagree on the principles adopted, a method should be prescribed for arbitration, such that a suitable method is established. The use of colour coding for marking verification sets is deemed to be beneficial in the recording of calculation status.

CALCULATION TITLE SHEET

SENIOR ENGINEER	CODE
CLIENT	
PROJECT TITLE	No OF SHEETS
PROJECT No.	VERIFIER
CONTENT	DATE
	SENIOR ENGINEER DATE
	APPROVING ENGINEER DATE
PREPARED BY DATE	

STATUS

S SCHEMATIC   W WORKING   T TENDER
P PRELIMINARY   O OTHER

COMPUTER ANALYSIS

YES [ ]   NO [ ]

ISSUE RECORD

REVISIONS

REV.	STATUS	DATE	DESCRIPTION	PAGE	BY	VER'D	APP'D

COMMENTS

Continuation Sheet Attached Y/N

Figure 6.4   Example of a calculation title sheet

Document issues

Engineering analysis and calculations, represent highly quality sensitive information and your procedure should, therefore, ensure that the detailed drawings produced are always prepared to the appropriate calculation revision. It should clearly state how the design information is processed to the detailer and that the construction drawings are verified to demonstrate compliance with the calculation. Clear records should also be evident on all issues, at each revision, and to which parties, for each set of calculations. This may well be achieved by the combined utilization of the title sheet, referred to previously, and a comprehensive transmittal sheet system.

Change control

Variations to input data occur regularly in the construction industry. Your procedure will need to recognize the sources of potential variations and to detail the manner in which those variations are accommodated within the design process.

This is often an area of confusion, and developing your system gives you a unique opportunity to analyse your procedures for receipt, assessment, processing and recording of variations. A comprehensive system in this respect will be of great benefit in ensuring that the real cost of design is reflected in the fee basis.

Feedback

Flexibility to accommodate varying site conditions is essential in the efficient programming of construction works. However, your system must ensure that such variations are accommodated in the design process and that the ultimate integrity of the structure is not compromised. Standard site report forms and concession notices are extremely useful in accommodating this requirement.

Preparation of detail drawings (2.6)

Procedures for the preparation of drawings already exist in most consulting practices. Formalizing these procedures should be a relatively straightforward task. The following notes can be adapted to suit the requirements of the individual organization:

1. Formal quality assurance requires that all drawings are prepared in accordance with recognized standards. These should be specified in the system manual.
2. Each drawing sheet should contain sufficient information for the drawing history to be reviewed; it must show the drawing to have been verified and approved at each revision.
3. Verification of drawings, as in the case of calculations, should be visible and, again, the use of colour codings by the verifier is useful. Record copies of verification drawings should be kept to ensure that audit of the process is possible. Drawing status should be clearly identified according to a system established throughout the company.
4. The drawing history record, maintained in a thorough and established manner, should include the following information as a minimum:

    *   date of commencement of project
    *   date from PQP for submission of working details
    *   revision of most recent issue
    *   number of bar bending schedule sheets attached to the drawing (where appropriate)
    *   issue number and date
    *   number of prints and destination at each issue.

Contract cost measurement (2.7)

This is an area of construction management that does not necessarily have an impact on the ultimate quality of the work. However, as quality assurance is a total management system, it embraces all aspects of construction management. Typical of the nature of the procedure for contract cost measurement would be as follows:

*   methods by which material take-off is verified to an accuracy sufficient for its purpose

- methods by which unit rates are established and updated
- methods by which bills of quantities are produced and verified
- methods by which tenders for construction works are assessed
- the procedure for performing interim and final certification in accordance with the various forms of contract and the verification of the final sums
- methods and responsibility for reporting the overall cost of the project to the client on a regular basis.

Control of records (2.8)

We have already discussed the necessity for comprehensive records for good contract management. The maintenance of such records in a standard format is essential to ensure continuity and retrievability at all times. For each project, only one set of records should be maintained in the project files, and the control procedure should ensure that these are comprehensive and current.

The following list would be typical for a design office:

- project correspondence – incoming/outgoing
- reports, specifications and manuals
- design calculations and verification copies
- computer record sheets, calculations and output
- drawing copies and verification sets
- bar bending schedules and material lists
- document transmittal sheets – incoming/outgoing
- drawing history and registers
- client and incoming drawing index
- confirmation of verbal instructions
- meeting records
- site visit reports.

Your procedure needs to identify how each file is titled and structured, so that all members of the design team and company staff find it easy to use. The procedure should also specify which records are to be kept in the archives following completion of the project, in what form and for how long.

Preparation of specifications (2.9)

Inadequate and/or incorrect specifications of the materials and workmanship for construction have been identified as a major problem in obtaining good quality construction. Your procedure will, therefore, need to identify the methods by which specifications are produced and verified in a thorough manner.

Control of the library (2.10)

Establishment of a technical library within the consultancy group is a requirement of the ISO standard for quality assurance. The following list is typical for civil and structural engineering design:

- specifications, British standards and codes of practice
- relevant professional, industry and international codes and specifications
- relevant product data, including British Board of Agrément documents
- new materials and their application
- design techniques
- defects and failure reports (including analysis and rectification, eg Building Research Establishment (BRE) documents)
- general technical information relating to civil, structural and other engineering disciplines relevant to the design activities for which quality assurance is applied.

In addition to the above, your procedure will need to ensure that all obsolete, superseded information is identified and removed or separated from current publications.

Concessions and corrective actions (2.11)

We discussed the necessity for accommodating site problems within the design process in the section on 'Design control'. It may, however, be sensible to incorporate a procedure within your system to accommodate the necessity for concessions and corrective actions throughout the process of contract document production.

Clearly, your procedure would need to specify how such variations and corrective actions are received, analysed, processed through the whole design and formally recorded.

Review of the QA system (2.12)

This is common to all sectors of the industry and has been covered already (pp. 58–62).

Placing contracts (2.13)

In order to comply with the requirements of formal Quality Assurance, a procedure will need to be established to ensure that all tenderers and subcontractors have sufficient experience and ability to carry out the work described in the tender documents. You may specify that evidence of Quality Assurance is required as prequalification to tender or, in the absence of such information, you may undertake to ensure that the working practices of the tenderers are adequate to provide the necessary quality of work.

Writing and controlling work instructions (2.14)

If it is your practice to issue notices detailing methods by which certain activities are undertaken, a procedure will be necessary to ensure that such instructions are prepared in an agreed format, that they are verified and that their distribution, amendment and withdrawal are controlled. The use of such instructions is attractive in the case where full procedures would otherwise be required.

Control of subcontract services (2.15)

Most practices, from time to time, employ the services of professional advisers to augment their project teams. A consulting engineer may, for instance, need help preparing construction cost estimates. A procedure is needed to ensure that such professional services meet the requirements of your quality system. Clearly, this will extend to the selection process, and advisers are often asked to comply with specific elements of your quality system, rather than evidence of their own quality procedures being insisted upon.

In addition to the above, it will be necessary to extend the audit process and design review to all subcontractors employed on your projects.

Equipment calibration (2.16)

A procedure will be required to demonstrate that all equipment utilized to produce sensitive information is calibrated, on a regular basis, within prescribed tolerances. In order to ensure that this discipline is adhered to, it may be cost effective to subcontract such calibration testing on a term contract.

Document issues (2.17)

The use of transmittal sheets is common within the construction industry. The extent and manner in which they are used, however, may not comply with the requirements of formal quality assurance. For example, it is common for such forms only to be used for the issue of working drawings. However, formal QA requires that they are used to transmit all quality sensitive documentation, including calculations, reports and specifications. In addition, formal quality assurance requires evidence that the information has been received by the appropriate individual, and colour coded triplicate forms are very useful here. Requesting the recipient to sign and date a copy of the form and return it to the originator ensures that evidence of receipt is established at the source.

Use of computers (2.18)

The use of computers in the design process is becoming more prevalent in all aspects of project development. If applicable, therefore, a procedure should be established to ensure the following:

- software is verified and validated for the required application
- modelling and formulation of analysis is verified by alternative means
- results are interpreted correctly and reconciled with inputs
- results are documented and processed to the design team in a thorough manner
- all members of staff utilizing the hardware/software are trained and competent.

The extent to which this procedure is developed will depend entirely on the sophistication of hardware and software to be utilized. Clearly, the use of finite element suites requires considerably higher levels of Quality Assurance than does the application of spreadsheets.

It is worth noting that bought-in software is not exempt from the requirements, and when selecting software it is prudent to request verification packages which can be run in-house, in order to avoid the expense of developing verification examples.

Contract management (2.19)

This procedure will cover all those activities carried out, following the completion of working drawings, to ensure that the works are constructed in accordance with the contract documents. The extent of involvement of the consultant in this activity will depend upon the nature of the contract documents and the extent of the works. The procedure will need to address the quality checks required to ensure that the materials and workmanship comply with the documents, and in addition that all necessary feedback to the design office is effected to ensure that the design remains competent.

The implementation of Quality Assurance across the whole spectrum of construction activities places upon the contractor an obligation to provide documentation to substantiate the quality of the works. This procedure may, therefore, concentrate upon ensuring that the contractor both supplies, in the first instance, and then complies with the necessary quality related documents.

Contract completion (2.20)

This procedure is likely to embrace many of the requirements previously stated in the above procedures. It may take the form of a checklist of requirements specified elsewhere. However, it may also be appropriate to require within the procedure a debriefing meeting to establish the lessons learnt throughout the period of the project. The procedure would state how these lessons would be recorded and fed back into the design process to ensure that errors are not repeated on future projects.

Using the quality system

The development and implementation of a Quality System can be time-consuming and expensive. So when a successful system is established, it should be used to every possible advantage. Your Quality System is a total management system and will address the full spectrum of company activity. This includes company promotion, tendering, project management and internal management training.

Company promotion

One obvious way of benefiting from a formal quality system is to apply for third-party certification.

There can be little doubt that a registration symbol on your letterhead will greatly enhance your credibility. In addition, such certification may permit inclusion on select lists that would otherwise be impossible. It is sensible, therefore, to account for the costs of certification against the company promotion budget rather than specific project cost centres; a 'registered firm' symbol and the associated audits will do little to enhance your ability to produce quality assured work, but certification could be the best investment in company promotion which you could make.

Tendering

There are two fundamental aspects to preparing a good tender.

- establishing client confidence in your ability to perform the work
- establishing the correct costs for the work.

With regard to establishing credibility, your quality system can offer a big advantage. The preparation of a procedure for PQPs establishes the format for control management of a project. Why not utilize this format for the preparation of tenders?

If you submit a PQP with the tender documents, you demonstrate to the client that you know his business, that you have clearly studied and understood the requirements of the contract and that you have the management control system to ensure that the work is carried out in a thorough and efficient manner. What better way of establishing your credibility?

Similarly, your quality system will almost certainly provide the management disciplines necessary to establish accurately the workscope associated with the tender, and thus to estimate the overall fee required. Senior management will then be in the fortunate position of reviewing the market and carrying out risk appraisal with confidence.

Fee tendering is fraught with problems and dangers, but a good quality system will do much to reduce the risks and enhance your ability to win the tenders and achieve profits and company growth.

Project management

The successful completion of a commission depends on the ability to start it in a coherent manner. The development of a project quality plan establishes, at a very early stage in the project, the control management disciplines necessary for successful completion. It acts as a checklist for the project manager to ensure that lines of communication, design basis and programme of work are clearly established. The PQP encourages a coherent approach by all members of the team which gives confidence to the client and everybody associated with the project.

Formal design and progress reviews act as a safety valve throughout the period of the project, to ensure that the requirements of the PQP are being met. As stated previously, they are great vehicles for training junior members of the team and keeping senior managers of the company visibly involved in the design process.

The establishment of an audit schedule also acts as a stimulant within the design team and ensures that the requirements of the PQP are being met. One of the most surprising aspects of establishing a formal quality system is that the audit is not regarded, by most members of staff, as a necessary evil. On the contrary, project teams tend to look forward to the audit as a challenge, and take great pride in receiving a good audit report. There is no doubt that this acts as a great morale booster and should not be overlooked as a principal benefit of establishing a quality system.

7 Quality Assurance and the Construction Contractor

Let us first examine who the Construction Contractor is. There are three main types of contractor, interfacing with the client in slightly different ways:

1. *Turnkey contractor*. Here the client is involved with a single organization which is responsible for the design and construction management of the whole project. The contractor may provide the design services himself or he may form a consortium or

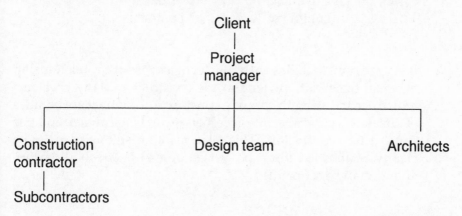

Client
|
Project
manager
|

Construction
contractor
|
Subcontractors

Design team

Architects

partnership with a design organization for the duration of the project.

2. *Independent contractors*. In this case the contractor or contractors are not working directly for the client/owner. They report through a Resident Engineer/Project Manager/Architect/Consultant Engineer to the client or owner.

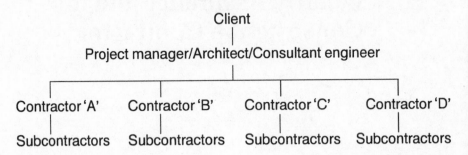

3. *Construction management*. For this type of contract the constructor liaises directly with the client/owner and is responsible to him for the management and construction activities, but not design.

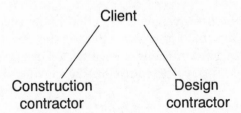

It is important to define at the outset of a contract the relationship between the client/owner and the constructor. Many problems which occur during the life of a construction contract are because of breakdowns at the interface between client/owner and the construction contractors. This relationship therefore needs to be clearly established from the outset, agreed between all parties and formally documented.

The contractor's responsibilities

The main responsibilities of the contractor do not differ greatly from one type to another. They include:

- setting up a contract team of experienced personnel
- selecting and appointing subcontractors
- planning for:
  a) resources
  b) equipment purchase
  c) materials purchase
  d) inspection and test methods
  e) preparation of procedures, instructions and method statements
  f) the control of issue of materials to site and on site
  g) setting up a site store
- controlling and monitoring all construction work
- reporting at regular intervals to the client
- preparing records
- complying with CDM 1994 requirements.

To satisfy the requirement of a quality assurance standard, whether it be ISO 9001 or ISO 9002, responsibilities need to be clearly defined both within the contractor's own organization and between the contractor and the client. A contractor who does not have a quality assurance system is unlikely to be able to satisfy the QA aware client who will sensibly seek evidence of QA capability before awarding a contract.

The tender

How does the contractor respond to a tender invitation and then set up a QA system for a construction project? The process commences with the QA aware client/owner writing QA conditions into the enquiry document for the contractor to respond to. Figure 7.1 shows the stages from the initial enquiry through to

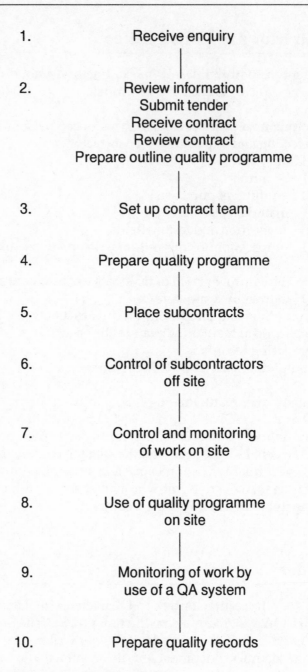

1.   Receive enquiry

2.   Review information
Submit tender
Receive contract
Review contract
Prepare outline quality programme

3.   Set up contract team

4.   Prepare quality programme

5.   Place subcontracts

6.   Control of subcontractors
off site

7.   Control and monitoring
of work on site

8.   Use of quality programme
on site

9.   Monitoring of work by
use of a QA system

10.   Prepare quality records

Figure 7.1   The contractor's QA activities

the final delivery of records to the client. We now examine each of these stages and look at the impact of the QA system on each.

Receive enquiry

The two documents usually received by the contractor from the client/owner at the tender stage of a contract with QA conditions are the technical specification and the QA conditions. The first defines what is expected in terms of:

- responsibilities
- interface with client
- qualifications and expertise of certain personnel
- technical requirements
- inspection and test requirements
- documentation, including drawings to be used
- records, and so on.

The Quality Assurance conditions request the contractor to provide documentary evidence that he has a management system in place to ensure that all the requirements of the technical specification will be carried out in a controlled manner. Also, that formal evidence will be generated during the various work phases as a record of the ongoing control and monitoring of the work. The contractor without a QA system will not be asked to tender because he cannot demonstrate, in a documented manner, his management controls. Enquiries and responses differ, but the following is typical.

Client's requirement	*Contractor's response*
1. The contractor must operate a QA system of not less than ISO 9002	1. Confirms in writing that a QA system to meet the requirements of ISO 9002 is operating
2. Submit documentary evidence that a QA system of not less than ISO 9002 is in operation	2. Submits a copy of the company's QA manual with a list of supporting procedures *or* sends evidence of approval of QA system by a recognized

Client's requirement	*Contractor's response*
	auditing authority *or* submits a list of clients where contracts have been successfully completed using a QA system
3. Submit an outline quality programme	3. Submits a typical organizational chart for a construction contract + typical responsibilities + a proposed audit schedule + a list of the company's procedures which would be applicable for that package of work

Other requirements which the contractor would have to satisfy, are likely to include:

● having a health and safety policy and recognition of CDM regulations
● having the technical capability and resources to carry out the contract
● having the financial stability, that is, likely to be in business at the end of the contract
● submission of a tender package with an acceptable price and timescale for completion.

A QA capability helps to satisfy these requirements. A QA system, for instance, requires training of personnel and keeping records of such training. It therefore helps to demonstrate that the contractor employs personnel with the necessary background experience and training to meet the requirements of the construction package. Another requirement of a QA system is to carry out contract reviews. The first contract review would be carried out at the tender stage by the contractor to assess all the technical, H & S and commercial implications of the enquiry. It would involve representatives from all departments preparing information for the tender. The formality of carrying out this contract review helps make the pricing and programming information to be included in the tender more realistic.

This initial contract review would cover the following:

- content of technical specification, including
  - a) special material, long delivery and so on
  - b) documentation
  - c) records
  - d) inspection and surveillance
  - e) resolving any queries with client/owner
- remoteness of site
- scope of work
- H & S requirements
- quality assurance requirements
- skills required
- human resources
- work to be subcontracted
- purchase of materials and equipment and so on.

Contract review

If the tender is successful, and a contract is received, the next stage is a further review which is an extension of that carried out at tender stage. It would include the following:

- checking to see if the contract document differs from the tender requirements
- resolving any differences with client/owner
- preparation of quality programme and quality plans
- setting up a contract team
- appointing subcontractors.

The contract reviews are attended by the relevant disciplines: Purchasing Manager, Site Manager/Resident Engineer, Quality Surveyor, Planner, Quality Manager/person responsible for QA, and so on.

Minutes of the contract review are kept as a record of the meetings. Any subsequent review meetings taking place as a result of scope of work changes are documented in a similar fashion. The objectives of contract reviews are to ensure that all aspects of the contract/tender are considered by the right people prior to work commencing, and if necessary discussed with the client to ensure everyone is working with the same information.

Setting up the contract team

Setting up a contract team is much the same process whether or not quality assurance is a contractual requirement. One important difference is that the organization and responsibilities of key staff must be formally defined, including that of the Quality Assurance Engineer. The information will form part of the Quality Programme and will include details of the interface with the client/owner (see Figure 7.2).

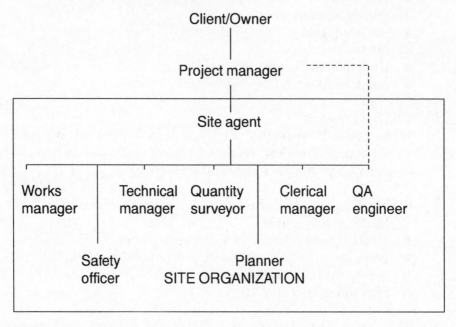

Figure 7.2   Organization and reponsibilities in a typical contract team

The project manager is responsible to client/owner for:

- the full execution of all construction work in accordance with the contract
- the administration of the contract in accordance with the approved quality programme
- the selection and management of subcontractors
- the preparation of temporary works designs and the safe completion of work
- preparation of monthly progress reports
- appointment of technical staff.

The site agent is responsible to the project manager for:

- organizing and managing all construction activities on site
- resolving dispute and technical queries
- financial control
- ensuring that all QA requirements are complied with
- ensuring that all construction activities comply with the technical specification and other contractual conditions
- labour relations.

The works manager is responsible to the site agent for:

- the recruitment, development and supervision of the construction labour force
- the control over plant and equipment used during the contract
- the supervision of quality and workmanship during the construction process.

The Quality Assurance engineer is responsible to the site agent (with functional supervision from the project manager) for:

- ensuring the Quality Assurance Programme is available on site and in operation
- carrying out induction training for staff in applying the Quality Programme
- preparation of Quality Plans
- monitoring work against Quality Plans

- monitoring work of subcontractors
- approving subcontractors' Quality Plans
- reviewing QA documents from subcontractors
- ensuring that audits are carried out to the agreed schedule and discrepancies resolved in a timely manner.

The above responsibilities are typical of a contract team as outlined in a Quality Programme. The make-up of a contract team varies greatly according to the size and complexity of a construction project, and job times may vary from one organization to another.

Preparing the Quality Programme

The Quality Programme/Project Quality Plan is the key Quality Assurance document. It is in essence a contract/project related quality assurance manual. Part of the document may have been prepared at the tender stage if the client had requested an outline Quality Programme. This Quality Programme will describe the management controls that are going to be implemented during the life of the contract. The contents of a typical Quality Programme are as follows:

- statement of the contractor's policy to carry out activities in accordance with a formal Quality Assurance system to the requirements of the applicable Standard as defined in the Quality Programme and supporting documentation. The statement confirms that the programme's contents are mandatory on all staff, and it is signed by the project manager
- table of contents
- amendment sheet
- abbreviations and definitions
- organization and responsibilities
- statement of project-scope
- procedures – these may be standard procedures taken from the company's Quality Assurance manual, supplemented by special procedures imposed by the client and procedures adopted to suit the client

- list of instructions, methods statements and inspection check-lists
- list, together with copies, of standard forms, such as defect notice, technical query form, concession note, inspection certificate, test certificate, stores release note and so on
- audit schedule, detailing the audits that will be carried out during the life of the contract to monitor both the contractor's and subcontractors' work. For an example of an audit schedule see Figure 7.3.

The Quality Programme will be submitted to the client for approval and work cannot generally commence until it has been approved.

There are circumstances in which not all documentation forming part of the quality programme can be made available prior to the contract commencing. The following documents, however, should be approved prior to that phase of work commencing:

- work instructions
- method statements
- inspection checklists
- quality plans.

Placing subcontracts

A contractor operating a Quality Assurance System is required to control the subcontractors. It is not sufficient to select subcontractors on the basis of price and delivery alone; the contractor must also ensure that they meet quality and H & S requirements.

Records of acceptable subcontractors are usually available in the form of a 'preferred list'. Information concerning subcontractors' Quality Systems can be obtained from a number of sources, for example:

- known history of the subcontractor, based upon previous experience

QA AUDIT SCHEDULE

PROJECT No. 737			
ISSUE 1	DATE 10/1/94		CLIENT – MANCUSTER SEWAGE PLANT

YEAR	1994												1995									
MONTH	J	F	M	A	M	J	J	A	S	O	N	D	J	F	M	A	M	J	J	A	S	O
DEPT/AREA/PROCEDURE																						
CONTRACTS OFFICE (PURCHASING DATA)	▓	▓																				
MATERIAL CONTROL			▓																			
CALIBRATION SYSTEM				▓	▓																	
TEST LABORATORY						▓	▓															
CONSTRUCTION					▓	▓			▓		▓	▓	▓	▓	▓	▓	▓	▓	▓		▓	
SITE OFFICE					▓																	
STORES					▓																	
ETC.																						

MANAGEMENT REVIEW

PREPARED BY:-   APPROVED BY:-

Figure 7.3   Example of a QA audit schedule

122

- certifications that the subcontractors possess, like Agrément Board Certification and CARES scheme
- carrying out of a quality audit of the subcontractor.

When choosing a subcontractor or supplier of materials, attention must be paid to any quality assurance requirement. Selecting the QA level is carried out in the same way as defined on pp. 63–4. According to the type of work being subcontracted, the QA standard imposed on the subcontractor/supplier will be ISO 9002, 9003 or commercial (that is, no formal evidence of QA required). Any QA requirements imposed on the subcontractor/supplier will be in addition to the technical specification and may be supplemented by a text amplifying the QA clause. An example of this text for a contract requiring a system of ISO 9002 is as follows:

For the duration of this Contract the Contractor shall operate to a Quality Assurance System not less than ISO 9002. In addition, one month prior to commencing manufacture/installation he shall submit the following documents to the Resident Engineer for approval:

- Quality Plan(s)
- Inspection Checklists
- Method Statements
- A copy of Quality Assurance Manual/Quality Assurance Programme to be used on this Contract if a copy is not already held.
  In addition to the above requirements the Contractor shall use the attached standard forms where applicable:
- Form 367 Technical Query Form
- Form 368 Concession Application Form
- Form 369 Defect Note (Remedial Work Form).

The above requirements can form part of the standard QA conditions for subcontracted work both on and off site.

Control of subcontractors (off site)

Because manufacturing operations (such as fabrication) are easier to control in a workshop environment, there is a move towards a modular type of construction, with assemblies being delivered to site.

The drawback to the system of off-site manufacture is that the main construction contractor may have to spend time and money monitoring this work in locations remote from the construction site. This is one of the benefits of choosing subcontractors and suppliers with good in-house management controls, that is, quality assurance systems.

Having confidence in the QA system of suppliers and sub-contractors allows for 'management by exception'. In other words, rather than making regular visits to the subcontractors'/suppliers' premises to monitor the work, you rely on their management control for assurance that the product or service will be delivered on time, on programme and to specification.

For certain types of work, especially if 'in process' controls are required, the contractor will wish to carry out a monitoring role. This is where the Detailed Quality Plan document plays an important part.

Detailed Quality Plan (DQP)

This type of Quality Plan which is going to be used to control a discrete package of work, should not be confused with Project Quality Plans/Quality Programmes which are project related quality systems.

This DQP is requested as a contractual requirement because it identifies the controls, responsibilities and evidence of satisfactory task completion. It fulfils some very important functions in the producing of work to meet specified requirements.

1. It demands thought about the requirements of the job, that is, it forces the subcontractor/supplier to read the specification and other key documents and then present in the form of a Quality Plan the means by which he is going to control the job.

2. It acts as a monitoring vehicle for stage-by-stage verification as work proceeds and thereby ensures that all planned actions are properly completed.
3. Because the contractor approves the Quality Plan prior to work commencing, it allows him to check whether the subcontractor/supplier has interpreted the specification correctly. There is also provision on the DQP for the contractor to identify any 'hold points' or 'witness points' where he wishes to monitor the progress of the job.
4. It provides a record at the end of the job of work integrity because, prior to signing off the Quality Plan, the contractor ensures that:
    ● all inspections/tests have been carried out
    ● all records have been raised
    ● all signatures have been completed.

Figures 7.3A to 7.3G (pp. 133–40) represent a typical specification and the ensuing DQP (Figure 7.4) submitted by a manufacturer of a precast concrete block where a requirement of ISO 9002 plus a quality plan was part of the contractual conditions.

Extended use of Detailed Quality Plan
It is common practice now to use the DQP format as a monitoring document to cover all stages of a project. This applies particularly to large complex projects where it is important that key aspects of the specification are identified and catered for by procedures, instructions or method statements and records of compliance produced and distributed. A typical example is shown in Figure 7.5 (p. 142) for a Design and Build Contract.

Many activities on this version of the DQP are 'generic', in other words are carried out repeatedly throughout a contract. In these cases (for example, Preparation of Method Statement, item 7) this activity can only be signed off on completion of all activities of that type. This may mean signing off some activities at the end of the contract.

In the example shown the Principal Contractor has appointed a Design Contractor nominated by the Client. The front page of this DQP would be the same as Figure 7.3C.

The first operations 1 to 9 are shown in the example but this document would probably extend to four or five pages to cover all key activities on the contract, finishing with the handover and warranty.

(A DQP would only be requested when the subcontractor's/supplier's package of work demands in-process controls, inter-stage inspections and so on.) For types of product not demanding evidence of in-process control, the contractor may choose to monitor the work by goods inward inspection.

Goods inwards inspection

Purchasing documentation will have clearly described the product ordered, including the type or grade, part number and the specification, drawing number or other technical require-ments, as applicable. The contractor's goods inwards inspection department will inspect the product and accompanying docu-mentation. Evidence of compliance which accompanies the product may be:

● inspection certificate
● test certificate
● material certificate
● certificate of compliance.

or whatever was demanded in the purchasing documentation. According to the type and quantity of the product being received, inspection may comprise:

● visual inspection for damage
● sample inspection of a batch
● 100 per cent inspection
● inspection for quantity only.

The degree of inspection carried out to monitor the purchased product will depend upon:

● confidence in the supplier's ability to deliver a satisfactory product

- the importance of the item (that is, could it stop construction?)
- ease of replacement if found wrong at a later stage.

The message to the construction contractor is that choosing the right subcontractor at the tender stage, even if it costs a little more, saves time and money at later stages because less monitoring of the supplier work will be required.

Control of subcontractors (on site)

Different controls can be applied, depending upon the nature of the work being carried out. Let's take, as an example, a client who is QA aware. This has been demonstrated by a request for a Quality Programme from the contractor for approval prior to construction commencing. The client/owner might also request, as part of the contractual conditions, DQPs for certain important work packages, such as:

- placing of concrete for foundations
- piling
- reinforced concrete structures
- erection of formwork
- laying of damp-proof membranes
- major structural steelwork.

In many cases, like pouring of concrete, a generic DQP could be used over and over again, with specific identification of the area, drawing number, building number and so on being added for each occasion.

Figure 7.4 (p. 141) shows a typical example of a DQP for placing concrete.

For other types of work it may not be necessary to produce a DQP, but records will still have to be produced to demonstrate control and verification of the work. Typical documentation would be procedures, instructions and method statements backed up by inspection checklists, inspection reports and test reports.

Use of the Quality Programme on site

For the on site work, the contractor's Quality Programme comes into full effect. It has been approved by the client/owner prior to work commencing, and represents a definitive statement of the management system that will be operated for the duration of the contract. The key aspects of the Quality Programme are:

● the organization and responsibilities
● the procedures, instructions/method statements covering the important construction and installation activities.

A number of other procedures will also be contained in the quality programme covering the other aspects of ISO 9002 which is the controlling standard for this contract, including:

ISO 9002 para	*Procedure Title*
4.5 Document control	The control and distribution of site documentation
	The control and distribution of documents to client
4.6 Purchasing	Purchasing of material and equipment for site use
	Verification of purchased material and equipment
	Monitoring of subcontractors/ suppliers
4.7 Control of customer-supplied product	Receipt and control of client supplied equipment and material
4.8 Product identification and traceability	Construction identification and traceability
4.10 Inspection and testing	Preparation of DQPs and inspection checklists
4.11 Control of inspection measuring and test equipment	Identification of equipment within the calibration system

4.12	Inspection and test status	Documenting the inspection status of the construction
4.13	Control of non-conforming product	Identification, documentation segregation and disposition of non-conforming product
4.14	Corrective and preventive action	Non-conformance review meeting
4.15	Handling storage, packaging preservation and delivery	Stores control
4.16	Control of quality records	Identification, collation, indexing, filing and storage of site records
4.17	Internal quality audits	Internal auditing
4.18	Training	Identification of training needs Preparation of training programme Training records
4.20	Statistical techniques	Preparation and use of sampling plans

Monitoring of work by use of a QA system

Having a Quality Programme is not a guarantee that the client will receive the constructed plant to time, programme and specification. The Quality Programme is a means to an end, and it relies on qualified, trained staff operating the system. In other words, you can have the best system in the world but it is *people* who make things happen. It is therefore important, when the quality programme is implemented on a contract, that adequate training is given to all personnel to make them aware of the procedures and the reason for having the system.

Once the contract is running, effectiveness of the quality programme needs to be monitored. Work in progress and any scope changes will be monitored by the procedural system which will probably include:

- use of DQPs
- use of inspection checklists
- progress meetings with client
- review of technical queries
- review of concessions
- review of non-conformances and defect work.

The contractor operating a Quality Programme to ISO 9002 must also carry out two other important types of review:

- management reviews
- internal audits.

The schedule for these should be agreed by the client at the time of approving the quality programme. See Figure 7.3 for a typical example of an audit schedule for a construction contract. This shows the management review as an annual event, and the audits scheduled to take place during the work phase in question. Records of the management reviews and internal audits are available as evidence that the system is being monitored, and that corrective actions have taken place as a result of problems being found.

Preparation of quality records

At the end of the contract the contractor will need to produce all the quality records required by the client and also the ones he will be responsible for keeping for a designated period of time. The contractor would be well advised to collect the records during the period of the contract. Not only does this prevent a mad rush at the end of the job, but it also gives confidence to the client, who can see records being handled in a controlled manner.

One advantage of working to a QA system is that work is carried out in a logical controlled manner, and from the start of the contract it should be clear which records are required, and when. Because there are now many ways of storing records, it is a good idea to agree with the client at the tender stage how the as-built records are required, for example:

- hard copies
- computer disks
- microfilm copies
- combination of the three.

The construction contractor will have to decide what records to keep and for what duration. This is both for his own record, if problems occur in the future, and also to satisfy any product liability legislation or CDM regulations.

The following is an extract from the specification placed by B & P Ltd on the Former Concrete Block Company (FCBC)

<u>Specification XYZ 623A for Precast Concrete Blocks.</u>

For the duration of this contract a Quality Assurance System of not less than ISO 9002 shall apply.

At least six weeks prior to manufacture commencing, a Quality Plan shall be submitted to B & P Ltd for approval.

The Quality Plan shall identify those drawings, documents, etc which are to be submitted to B & P Ltd for approval and shall detail specific operations, inspections and tests as deemed necessary by the contractor to give adequate control over the process. The contractor shall also add references to his own internal procedures.

B & P Ltd reserve the right to amend the Quality Plan and insert 'hold points' on specific activities.

<u>Drawings</u>
A mould drawing shall be submitted to B & P Ltd for approval prior to manufacture commencing. The drawing shall meet the requirements of the attached:

    CB 6372 A Arrangement of precast concrete blocks
    CB 6373 A Detail sections of precast concrete blocks
    CB 6374 B Arrangement and location of reinforcement steel.

Figure 7.3A   Typical contract requirements

Technical Queries
All technical queries shall be submitted on the B & P Ltd, T Q form CC 2093 A and agreed prior to work commencing.

Concessions
All applications for concessions shall be submitted on B & P Ltd Concession Form CC 2094 B and items shall not be dispatched prior to approval being given.

Repairs
Repairs shall be requested via the Concession Application forms.

Subcontractors/Suppliers
A list of proposed manufacturers and sources of supply for the listed items shall be submitted for approval prior to enquiries being sent out:-

> Cement
> Aggregates – Coarse
> Aggregates – Fine
> Admixes
> Curing agent
> Reinforcement bar (from a CARES approved supplier)
> Water

Mould Preparation
The procedure for mould preparation shall be submitted for approval prior to manufacture commencing.

Pre and Post Pour Check Sheets
Pre and post pour check sheets shall be submitted for approval prior to manufacture commencing.

Calibration
Measuring and test equipment shall be covered by the contractor's calibration system and records shall be available to verify same.

Figure 7.3A   (continued)

Inspection and Test
The results of slump tests, cube tests and dimensional checks shall be made available to B & P Ltd.

Documentation
A documentation package shall be gathered together prior to packing and dispatch, containing the following:

> Completed Quality Plan
> Material Certificates
> Inspection Certificates
> Batch Certificates
> Test Certificates
> Repair Certificates
> Approved Concessions
> Agreed TQs.

Storage
All materials and finished items shall be stored undercover in a dry store and identified in a manner to prevent loss, damage or deterioration.

Figure 7.3A   (concluded)

The following are extracts from the Former Concrete Block Company Q.A. Manual.

F.C.001	– Preparation of Quality Plans
F.C.002	– Procurement of materials and services
F.C.003	– Drawing preparation and approval
F.C.004	– Preparation and approval of calculations
F.C.005	– Calibration of instruments and test equipment
F.C.006	– Stores control
F.C.007	– Preparation of inspection and test procedures
F.C.008	– Pre and Post pour check lists
F.C.009	– Documentation packages/records
F.C.0010	– Packing and dispatch
F.C.0011	– Use of CARES approved suppliers
F.C.0012	– Concessions
F.C.0013	– Technical queries
F.C.0014	– Goods Receipt Inspection
F.C.0015	– Mould preparation
F.C.0016	– Operation of Batcher Plant
F.C.0017	– Preparation, curing and testing of concrete
F.C.0018	– Rectification of pre-cast blocks
F.C.0019	– Inspection of pre-cast blocks

Figure 7.3B   Former Concrete Block Company procedures/instructions

FORMER CONCRETE BLOCK Co.

QUALITY PLAN No.   F1070
JOB REF. OR DESCRIPTION: PROJECT-MANCUSTER
                                SEWAGE PLANT
CLIENT: B & P Ltd.
PLAN PREPARED BY: P SUMNER   DATE: 2/8/95 ISSUE A
PLAN APPROVED BY: B THORPE   DATE: 2/8/95
CLIENT APPROVED BY: DATE:

KEY DOCUMENTS

TITLE	AMEND'T	BY	DATE	APPROVED BY
DRG No. CB 6372 A				
DRG No. CB 6373 A				
DRG. No. CB 6374 A				
CODE N/A				
SPEC'N XYZ 623 A				

Figure 7.3C   Former Concrete Block Company DQP

PLAN COMPLETIONS

	SIGNED OFF	DATE
ALL DOCUMENTS ISSUED TO PLAN		
ALL VERIFICATIONS CARRIED OUT TO PLAN		
ALL TESTS COMPLETED TO PLAN		
ALL RECORDS PRODUCED TO PLAN		
CUSTOMER FINAL ACCEPTANCE		

PLAN CODES:

A – DOCUMENTATION SUBMISSION   C – RECORDS FOR CLIENT
F – FINAL INSPECTION   R – RECORDS FOR FCBC
H – HOLD POINT
I – INSPECTION 100%
P – SAMPLE INSPECTION   SHT. 1 OF 5

Figure 7.3C   (concluded)

OP. No.	OPERATION	REF. DOC'T	VERIFYING, CODE & SIG.						VERIFYING DOCUMENT	RECORDS					
			F.C.B.C.			CLIENT				F.C.B.C.			CLIENT		
			CODE	SIG.	DATE	CODE	SIG.	DATE		CODE	SIG.	DATE	CODE	SIG.	DATE
1	DOCUMENTATION SUBMISSIONS														
1.1	QUALITY PLAN	SPEC XYZ 623 FC 001	AH						APPROVED QUALITY PLAN	R					
1.2	TECHNICAL QUERIES	CC 2093 A FC 0013	A						TQ RESPONSE	R					
1.3	MOULD DRAWING AND PROCEDURE	CB 6372 CB 6373 CB 6374 FC 003 FC 0015	A						APPROVED DRAWING ABC607 & PROCEDURE FC 0015	R					
1.4	PRE AND POST POUR CHECK SHEETS	FC 008	A						APPROVED PRE & POST POUR CHECK SHEETS	R					
1.5	LIST OF PROPOSED MANUFACTURERS/ SOURCES OF SUPPLY FOR:- -CEMENT -AGGREGATES COARSE -AGGREGATES FINE -ADMIXES - CURING AGENT - REINFORCEMENT STEEL (CARES APPROVED SUPPLIED)	FC 002 AND FC 0011	A						AGREED LIST OF SOURCES OF SUPPLY	R					

QUALITY PLAN NO: F1070
PREPARED BY: P SUMNER
DATE: 2/8/95 ISSUE: A

Figure 7.3D Continuation Sheet No. 2 of 5

137

OP. No.	OPERATION	REF. DOC'T	VERIFYING, CODE & SIG. F.C.B.C. CODE	SIG.	DATE	CLIENT CODE	SIG.	DATE	VERIFYING DOCUMENT	RECORDS F.C.B.C. CODE	SIG.	DATE	CLIENT CODE	SIG.	DATE
2 2.1	PRE MANUFACTURE ISSUE CONTRACT DOCUMENT	FC 002 AGREE LIST OF SUPPLIERS	I						PURCHASE ORDERS CONTRACTS	R					
2.2	CHECK MATERIAL RECEIVED TO SAMPLING PLAN	FC 0014 PURCHASE ORDERS	P						MANUFACTURERS CERTS.	R					
2.3	CHECK STORAGE CONDITIONS	FC 006	I						STORES RECORDS	R					
2.4	CHECK CALIBRATION STATUS OF EQUIPMENT	FC 005	I						CALIBRATION RECORDS.	R					
2.5	PREPARE MOULD	FC 0015	I						INSPECTION CERT.	R					
2.6	PREPARE REINFORCEMENT		I							R					
2.7	ASSEMBLY MOULD AROUND REINFORCEMENT		I							R					

QUALITY PLAN NO: F1070
PREPARED BY: P SUMNER

Figure 7.3E   Continuation Sheet No. 3 of 5

138

OP. No.	OPERATION	REF. DOCT	VERIFYING, CODE & SIG.						VERIFYING DOCUMENT	RECORDS					
			F.C.B.C.			CLIENT				F.C.B.C.			CLIENT		
			CODE	SIG.	DATE	CODE	SIG.	DATE		CODE	SIG.	DATE	CODE	SIG.	DATE
3	PRE POUR CONTROLS AND INSPECTIONS														
3.1	CONCRETE BATCHING	PRE POUR CHECK SHEET FC 0017	I						BATCH CERT	R					
3.2	CONCRETE TESTING		I						TEST CERT	R					
3.3	PERMISSION TO POUR		I						RELEASE NOTE	R					
4	MANUFACTURE														
4.1	POUR CONCRETE		P						INSPECTION CERT	R					
4.2	CURE CONCRETE		P							R					
4.3	CHECK RESULTS & IDENTIFY TEST PIECES		I							R					
5	POST POUR CHECKS	POST POUR CHECK SHEET FC 0019													
5.1	STRIP MOULD		F						INSPECTION CERT	R					
5.2	LIFT AND STACK														
5.3	IDENTIFY														
5.4	INSPECT														

QUALITY PLAN NO: F1070
PREPARED BY: P SUMNER
DATE: 2/8/95 ISSUE: A

Figure 7.3F   Continuation Sheet No. 4 of 5

139

OP.	OPERATION	REF. DOCT	VERIFYING, CODE & SIG.							VERIFYING DOCUMENT	RECORDS						
			F.C.B.C.			CLIENT					F.C.B.C.				CLIENT		
No.			CODE	SIG.	DATE	CODE	SIG.	DATE			CODE	SIG.	DATE	CODE	SIG.	DATE
6 6.1	RECTIFICATIONS REPAIRS	FC 0018	I							INSPECTION CERT.	R					
6.2	CONCESSION APPLICATIONS	CC2094B FC 0012	A							APPROVED CONCESSION FORMS	R					
7	COMPILE DOCUMENTATION	FC 009	H							CLIENTS RELEASE NOTE	R					

QUALITY PLAN NO: F1070
PREPARED BY: P SUMNER
DATE: 2/8/95 ISSUE: A

Figure 7.3G   Continuation Sheet No. 5 of 5

140

CCC CONCRETING Co. QUALITY PLAN CLIENT: MANCUSTER SEWAGE Co. AREA: BLD 32 AREA 2G			CONTRACT No. Z 3756 A PAGE: 1 OF 1		
TITLE: PLACING CONCRETE FOR FOUNDATIONS			DATE: 2/8/95		
OP No.	OPERATION	CONTROL DOCUMENT	CODE	VERIFYING DOCUMENTS	SIGNATURE & DATE
1	ORDER CONCRETE	SPECIFICATION C 600	F	PURCHASE ORDER	
2	CHECK: –FORMWORK IS COMPLETE AND SAFE. –REINFORCEMENT IS IN PLACE – POUR IS CLEANED OUT AND IS SAFE FOR POURING –ANY SUPPORTS ARE INDEPENDENTLY CHECKED. –WATERBARS.	CHECK LIST C 30 A	F F F F.SE F	SIGNED OFF CHECKLIST	
3	GIVE PERMISSION TO POUR	PERMISSION TO POUR SHEET	SE	PERMISSION TO POUR SHEET	
4	ACCEPT CONCRETE FROM BATCHER PLANT	SPECIFICATION C 600	SE	DELIVERY TICKET	
5	SEND SAMPLE TO CONCRETE LAB.	CHECK LIST C 30 A, SAMPLE FROM C 20	F	TEST CERTIFICATE	
6	PLACE CONCRETE AND VIBRATE	METHOD STATEMENT M 562	CT	SIGNED OFF CHECK LIST	
7	VISUALLY INSPECT	CHECK LIST C 30 A	SE	SIGNED OFF CHECK LIST	
8	DEFECT NOTICE OR CONCESSION NOTES	PROCEDURE CC 42	F	AGREED CONCESSIONS AND REPAIRS	

RESPONSIBILITY CODES

F – FOREMAN
SE – SECTION MANAGER
CT – CONCRETING TEAM

	SITE ENGINEER	CLIENT
COMPLETION OF WORK	Sig. Date	Sig. Date
DOCUMENT COMPLETE	Sig. Date	Sig. Date

Figure 7.4   Detail Quality Plan for placing concrete

OP. No.	OPERATION	REF. DOC'T	VERIFYING, CODE & SIG.						VERIFYING DOCUMENT	RECORDS						COMMENTS
			CONTRACTOR			CLIENT				CONTRACTOR			CLIENT			
			CODE	DATE	SIG.	CODE	DATE	SIG.		CODE	DATE	SIG.	CODE	DATE	SIG.	
1	Contract Review	P03	1						Contract Form F02	R						
2	Prepare Quality Programme	P06	1 A			H			Quality Programme Form F06	R			C			Submit within 21 days of tender acceptance
3	Prepare Contract Programme		1 A			H			Contract Programme	R			C			
4	Appoint design contractor	P08	1 A						Contract	R						Nominated by Client
5	Submit design work for client acceptance	P11	1 A			H			Drawings Calculations	R			C			All drawings and calculations to be verified
6	Implement CDM requirements	Contract documents P18 H&S policy	A			H			Health & Safety Plan and Safety File	R			C			
7	Prepare method statements	P15	1 A			H			Approved Method Statements	R			C			
8	Submit testing and inspection schedule	P17	1 A			H			Approved testing and Inspection Schedule	R			C			
9	Appoint subcontractors	P22							Sub orders	R			C			Use approved suppliers' list
ETC																

Figure 7.5   Use of DQP as a monitoring document

8 QA Interfaces (Summary)

QA activities on a typical construction project

This flow chart identifies the key quality assurance activities which would take place during the life cycle of a typical construction project, from project initiation to handover.

CLIENT	DESIGN CONSULTANT	CONSTRUCTOR

Establish project brief/objectives/specification (include QA conditions) ──▶ Carry out Tender Review, prepare outline PQP, and submit

Accept outline PQP ◀──

Place contract ──▶ Set up project team, carry out contract review, prepare PQP

143

CLIENT	DESIGN CONSULTANT	CONSTRUCTOR
Approve PQP ◄───	Submit for approval	
Approve DQP ◄──►	Prepare DQP – if appropriate	
Approve key ◄──► drawings	Prepare drawings	
Approve ◄──► calculations	Prepare calculations carry out design reviews, prepare detailed specifications	
Monitor design consultant's activities by audit	Issue enquiries for ──► construction work including QA conditions	Carry out Tender review, prepare QA submission
	Carry out bid appraisal, assess QA submission ◄───	
	Place contract with QA conditions ──►	Carry out contract review Set up site team
	Approve PQP ◄──►	Prepare PQP and submit for approval

CLIENT	DESIGN CONSULTANT	CONSTRUCTOR
		Place subcontracts including QA condition where appropriate to work package. Include requirement for documentation submissions,
	Approve shop ◄──► drawings and other documents	approvals and records
		Receive DQPs from subcontractors for approval prior to work commencing
	Approve DQPs	Prepare DQPs for own work if required
	Project manage contract. Carry out progress meeting. Monitor work by inspection test and review of documentation	Place 'hold points' etc on DQPs to monitor work packages. Approve DQPs
		Monitor off-site work against DQPs
	Carry out audits to agreed schedules	

CLIENT	DESIGN CONSULTANT	CONSTRUCTOR
		Carry out goods inwards inspection to agreed procedure
		Control work on site against PQP, DQPs, inspection checklist, etc.
		Carry out audits on and off site to agreed audit schedule
		Generate records as construction proceeds
		Mark up drawing to as-built state
		Prepare handover packages and submit
Accept ◄——— documentation package	Check handover package and submit with design records ◄———	

NB   The above does not include for responsibilities arising as a result of the CDM (1994) regulations

9 Construction Design and Management (CDM)

From 31 March 1995 there has been a requirement for the construction industry to design and contract to meet the requirements of the CDM 1994 regulations.

Legal obligations are placed on everyone involved in the construction process, including the client, to provide for site safety at all stages of a project. This means that the contractor will no longer take sole responsibility. The client commissioning the project and other construction professionals such as consulting engineers, project managers, surveyors and architects will all now have to ensure that they meet their legal obligations.

Figure 9.1 illustrates the relationships between the various duty holders and the importance of the Health and Safety Plan which is the key link between the planning supervisor and the principal contractor. The plan has to be initiated by the planning supervisor, in liaison with others (such as the designer), and developed sufficiently to form part of the tender documentation for issue to the contractors.

The principal contractor awarded the contract should further develop the plan to cover the proposed site activities and the contractors involved. Construction cannot commence until the client

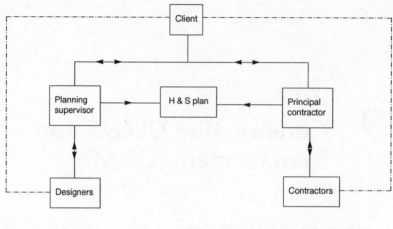

Indicates additional direct contractual links

Figure 9.1   Simple diagram showing key relationships between the principal parties as defined under the CDM Regulations

has assured himself that the plan has been prepared.

The above paragraphs embrace only a small part of the contents of the CDM regulations. The effect of CDM on a company's Quality Assurance system will almost certainly mean modification to existing organization charts to reflect the role of, or liaisons with, the planning supervisor, and so on.

For example:

Client

● Appointment of designer(s) for CDM work
● Appointment of planning supervisor
● Appointment of contractors for CDM work
● Assessment of H & S plan

Planning supervisor

● Preparations of pre-tender H & S plan
● Carrying out the planning supervisor's duties and responsibilities
● Compiling H & S file

Principal contractor

● Developing H & S plan
● Appointment and monitoring of contractors with respect to H & S plan requirements
● Training of staff and contractors and providing risk information

Contractors

● Interface with principal contractor on H & S plan requirements

Designer

● Interface with planning supervisor on pre-tender H & S plan ·
● Interface with client on H & S issues
● Risk assessment in design

Index

151